THE CARBON CRUNCH

THE
CARBON
CRUNCH

**How We're Getting Climate Change Wrong
– and How to Fix It**

DIETER HELM

YALE UNIVERSITY PRESS
NEW HAVEN AND LONDON

For information about this and other Yale University Press publications, please contact:

U.S. Office: sales.press@yale.edu   yalebooks.com
Europe Office: sales@yaleup.co.uk   www.yalebooks.co.uk

Set in Minion Pro by IDSUK (DataConnection) Ltd
Printed in Great Britain by TJ International Ltd, Padstow, Cornwall

Library of Congress Cataloging-in-Publication Data

Helm, Dieter.
    The carbon crunch: how we're getting climate change wrong—and how to fix it/Dieter Helm.
        p. cm.
    Includes bibliographical references and index.
    ISBN 978-0-300-18659-8 (hardback)
1. Energy policy. 2. Renewable energy sources. 3. Energy conservation. 4. Climatic changes—Prevention. 5. Greenhouse gas mitigation. I. Title.
    HD9502.A2H455 2012
    333.79—dc23

                                                                                      2012017386

A catalogue record for this book is available from the British Library.

10 9 8 7 6 5 4 3 2

2016 2015 2014 2013 2012

Contents

List of Figures and Tables

List of Abbreviations

AGR, advanced gas-cooled nuclear reactor

BRICS, Brazil, Russia, India, China, South Africa

CAFE, Corporate Average Fuel Economy

CCGT, combined cycle gas turbine

CCS, carbon capture and storage

CDM, Clean Development Mechanism

CEGB, Central Electricity Generating Board

CER, Certified Emissions Reduction

CND, Campaign for Nuclear Disarmament

CO, carbon monoxide

CO_2, carbon dioxide

CO_2e, carbon dioxide equivalence (i.e., including all greenhouse gases)

CSP, concentrated solar power

CTL, coal-to-liquids

EIA, US Energy Information Administration

EU ETS, European Union Emissions Trading Scheme

FGD, flue-gas desulphurization

FiT, feed-in tariff

GDP, Gross Domestic Product

GW, gigawatt

IAM, integrated assessment model

IEA, International Energy Agency

IEM, internal energy market

IMF, International Monetary Fund

IPCC, Intergovernmental Panel on Climate Change

kWh/d, kilowatt-hour per day

LCPD, EU Large Combustion Plant Directive

LNG, liquefied natural gas

MIT, Massachusetts Institute of Technology

MMBtu, million British thermal units

MW, megawatt

NGO, non-governmental organization

NOx, nitrogen oxides

NPV, net present value

OPEC, Organization of Petroleum Exporting Countries

ppm, parts per million

PV, photovoltaics

PWR, pressurized water nuclear reactor

R&D, research and development

REDD, Reducing Emissions from Deforestation and Forest Degradation

ROC, Renewable Obligation Certificate

SOx, sulphur oxides

SUV, sports utility vehicle

UEA, University of East Anglia

UNFCCC, United Nations Framework Convention on Climate Change

WTO, World Trade Organization

Preface and Acknowledgements

I have written this book because in almost a quarter of a century virtually nothing of substance has been achieved in addressing climate change. Despite a multitude of summits, speeches and commitments, and a host of energy policies, emissions keep going up. Not even a world recession has made much difference. The concentration of carbon in the atmosphere is well on the way to doubling from its pre-Industrial Revolution level, and as yet nothing much stands in the way of it trebling by the end of this century. It is hard to find any mainstream climate change scientist or economist who now believes that global warming can be limited to 2°C.

I have written extensively about various aspects of climate change and energy policy over a number of years, and in 2011 Phoebe Clapham at Yale University Press approached me and suggested that I bring my ideas together in an accessible book. In trying to achieve this, I have eschewed the academic practice of peppering the text with innumerable references, and (I hope) present my arguments in an approachable, non-academic format, whilst backing them up with endnotes. I have also avoided the temptation to provide reams of empirical evidence, partly in order to make the book tractable, but more importantly because climate and energy policy is not about precise predictions and forecasts of the costs of each and every technological option. Indeed, the model-driven approach is, as will be explained, one of the reasons why we have ended up in the present mess. The very complexity of the now enormous literature on climate change in the fields of science and economics has become a barrier to a wider understanding amongst those who are being asked to pay the bills.

I am very grateful to Phoebe, not only for suggesting this book, but for pushing me to complete it in a short time. She has also provided extensive comments on the drafts, and has been extremely supportive in every way.

Many people mistakenly believe that their ideas are original, when they are in fact derived from the work of others. In mid-career, I had the good fortune to meet Cameron Hepburn, with whom I have since collaborated on a number of papers and books. He has shaped and changed my views on many aspects of climate change, and although he should not be blamed for any of my many errors and misunderstandings, or even for the main thrust of the argument presented here, it is hard to think that I could have brought my ideas together without his astute questioning and encouragement.

Many others have had a hand in the shaping of my arguments. The late David Pearce, who pioneered the economic approach to environmental policy in the 1980s and 1990s, was a great influence on me and, like many, I miss him greatly. Other academics who have shaped my views include Colin Mayer, a sharp and insightful critic throughout my career, and Chris Allsopp, a colleague at New College.

Richard Green provided detailed and extremely helpful comments on an early draft. I am grateful too to Bob Hahn for his comments and our very useful and often highly entertaining conversations on both the politics and economics of climate change. Other influences include Alex Kemp, Giles Atkinson, David Newbery, Nicholas Stern, Paul Stevens, Rick van der Ploeg, Roger Scruton, David Victor, Stephen Glaister, Luis Correia da Silva, Helen Jenkins, Robin Smale, David MacKay, David Maddison, David Wiggins, Stephen Smith and Scott Barrett.

During 2011, I was a special adviser to the European Energy Commissioner, Günter Oettinger, and also chair of the Advisory Group on the EU 2050 Energy Roadmap. I saw at first-hand how Brussels works.

It was not always a pretty sight, notwithstanding the quality and integrity of many of those who toil away in the Energy Directorate – especially Philip Lowe. My experience of the conduct of the lobbyists – particularly the green politicians and green NGOs – was very disillusioning to someone who is both deeply concerned about green issues and who has been deeply pro-European, but it did help me to understand how European climate and energy policy has ended up the way it has, and why it is so difficult to change.

In Britain, I have served on various government energy advisory bodies for the last two decades: first on the Department of Trade and Industry's Energy Advisory Panel, followed by the Sustainable Energy Policy Advisory Board, then the Advisory Panel on Energy and Climate Security, and now the Energy Advisory Group of the Department of Energy and Climate Change. I also spent three years on the Prime Minister's Council for Science and Technology. In these roles and elsewhere, I have met with every British Secretary of State for energy – and their counterpart on the opposition bench – since 1980, and from some I have learnt a lot.

Of the many (roughly one per year), there have been some outstanding politicians. Nigel Lawson remains the towering figure behind the design and structure of the market-based approach, which he pioneered in the early 1980s. Tony Blair steered the Labour Party opposition through the privatization of the electricity industry at the end of the 1980s as Shadow Secretary of State. John Hutton and Tony Blair again confronted the sad state of energy policy that had developed by the middle of the first decade of this century, and Ed Miliband shaped the major reforms of the energy markets before the 2010 general election. I have learnt much from all of them – especially the difficulty of reconciling the politics of energy policy with academic theories and concepts.

As university economics departments have concentrated on mathematics and econometrics, and largely deserted more policy-orientated subjects like energy, this domain has been occupied by civil servants, regulators, advisers and those working for think tanks. Amongst the many impressive people, Anna Walker, Chris Bolt, Stephen Littlechild, Andrew Wright, Steve Hilton,

Ben Moxham, Simon Less, Steven Fries, David Kennedy, Adrian Gault, Simon Virley, Agata Hinc and Pawel Swieboda have all influenced my thinking. Ben Caldicott offered both encouragement and detailed comments, and I have greatly benefited from our discussions.

I have also had the benefit of extensive discussions and debates with those on the industrial side. Many people in the green movement believe that even talking to companies is a sin, especially if those companies are not in the renewables business. This belief is deeply misguided: it is companies, not governments, that build power stations, manufacture energy efficiency materials and deliver energy to customers, and the contempt too often piled upon anything which has a 'corporate' association is at best unhelpful and at worst very damaging. Inside energy companies I have met with very talented people, who often have a rigour of analysis that should be welcomed. Amongst the many who have either commented directly on draft chapters or influenced my thinking are Richard Ritchie, Paul Spence, Sara Vaughan, Christophe Bonnery, Denis Linford, Peter O'Shea, David Mannering, Rupert Steele, Richard Howard, Chris Anastasi, Barry Neville, Andrew Mennear, Peter Mather, Michael Rolls, Janine Freeman, Paul Whittaker and Robert Sorrell. None of these, or indeed any of the academics and policy people mentioned above, is responsible for my many errors, or should be assumed to agree with the arguments presented here.

In the academic and policy world, any academic who teaches bright young students knows that the bargain in a tutorial is very one-sided in favour of the teacher. Having to explain one's ideas and have them challenged by one's students has been a privilege, and there have been many New College students over the years who have inadvertently altered my views. In an Oxford college, colleagues come from the full range of academic disciplines, unlike in a department-based university system. In New College, I have greatly benefited from conversations with engineers, scientists, historians, philosophers and political scientists. The Warden, Curtis Price, and the Fellows have allowed me to develop a wide range of interests, with a generosity for which I am extremely grateful.

Putting together this book in a short period has been a major task, and I have had two crucial supports – Vicky Hibberd and Kerry Hughes. Vicky has tirelessly researched facts, charts, and papers, and provided an invaluable flow of inputs, whilst at the same time dealing with my scribbles. Kerry has done a brilliant job of copy-editing the book, and has sorted out the grammar, style and presentation of the material. I am very grateful to both of them.

Finally, all authors know how much their families suffer once a book gets a grip on them. In my case I have had the benefit not only of Sue Helm's tolerance, but also her comments on both the general arguments and the draft. The book is for her, and Oliver and Laura, in the hope that good sense will prevail, and that after a quarter of a century of largely futile efforts and costs, something will finally be done about climate change.

Introduction

Imagine a historian writing in 2050, or even 2100, about the late twentieth and early twenty-first centuries. No doubt the current Great Recession will be prominent, and compared with the Great Depression of the 1930s. The rise of China, the nuclear challenge of Iran, and the Arab Spring might all feature. But what will also stand out is that this was the first time that humans faced a major global environmental threat, and politicians started to grapple with the consequences. Rising global temperatures, almost certainly caused by human activity, threatened the very idea of economic growth, and if unchecked could wreak havoc upon not just humans but the rest of the natural world, radically altering habitats and accelerating extinctions.

The sheer scale of the challenge contrasts with the stuff of day-to-day politics: nothing like it – not even regional conflicts – demands so much in terms of international cooperation and the need to take serious account of the welfare of future generations. Our historian would need to compare these environmental challenges with the great disruptions in geological time rather than recent human history.

Our historian will of course have the benefit of hindsight, knowing how it turned out: whether it is much hotter when the history is written, how many species are left, and whether climate change has had all the terrible consequences some scientists predict. We, on the other hand, are condemned to live with the uncertainty. But we are not completely in the dark. We can already see how politicians and policy-makers have responded so far, whether they are rising to the challenge, and whether the magnitude of the problem has spurred leaders across the globe to develop a coherent strategy and achievable measures designed to address these challenges, whilst not disadvantaging individual states and sections of society.

In the first two decades of tackling the climate change problem, our chiefs cannot be accused of not trying. World leaders confronted climate change in the early 1990s, and two decades ago signed the UN Framework Convention on Climate Change (UNFCCC). They have made innumerable speeches, attended countless conferences, and spent a great deal of energy customers' and taxpayers' money since then. Sadly there is little to show for it. Two decades on, emissions continue to rise alarmingly. If anything they are accelerating. From a pre-Industrial Revolution concentration of around 275 parts per million (ppm) of carbon[1] in the atmosphere, we are rapidly approaching 400 ppm. Back in 1990, around 2 ppm were added annually. Now it is nearly 3 ppm. Scientists tell us that 400–450 ppm is roughly associated with 2°C warming.[2] If they are right, and if we go on as we are, it is going to be more – potentially much more – than 2°C.

These numbers really matter. Try to imagine being in the historian's shoes. What will the world be like if the temperature rises by, say, 3°C by 2100. It will be very different. The best places to live may be in the Arctic – Canada, Russia, Greenland and Alaska. Their pristine wildernesses may be filled with new cities and dense populations. Unfortunately the southern parts of the southern hemisphere offer little by way of habitable lands even in a warmer world – there is a lot of ocean, and the ice sheet over Antarctica will probably still be largely intact. The tropics will be very tropical. Low-lying lands will suffer from rising sea levels.

At 3°C, despite all the disruption, the world will still get by. At 6°C it is the stuff of nightmares – and scary films. It might even at these extremes start to flip into catastrophic climate change, as methane could be released in large quantities from the tundra, and feedback loops turn really nasty.

Then again, all this might not happen. It might be only 1°C – and up to 2°C is not all bad. Indeed, in many areas a bit of global warming might be a good thing: warmer winters, less heating, longer growing seasons, and lots of ice-trapped minerals now accessible. In many cases, things will probably get better before they get worse.

We are condemned to uncertainty. We do not have the historian's luxury of hindsight. But few doubt that the scale of the threat is worth dealing with. They may differ about the urgency and the prioritization of climate change over other challenges like our Great Recession, but the fact is that so far nothing much has been done.

Over the period since 1990, the growth of emissions has been the result of an unprecedented economic expansion, and one based on coal and China and population. A remarkable economic transformation has taken place. The US (and to a lesser extent Europe) embarked on a debt-fuelled spending spree, and as a result sucked in goods from China and other developing countries. The scale has been breathtaking. And it inevitably resulted in a spectacular bust. China, emerging from the deadly embrace of Mao, started growing, and for the two decades racked up around 7–10% growth in GDP per annum. At this rate, it has doubled in size every decade, so it is already four times bigger than it was in 1990. This growth has been energy- and carbon-intensive, and it has been fuelled overwhelmingly by burning coal – the dirtiest of all energy sources.

While all this has been going on, what did world leaders achieve? Our historian might reasonably start off with the causes of the human-induced emissions, and then see whether they addressed them. They haven't so far. Instead, in 1992, the focus was on setting caps on emissions in the key *developed* countries, in the hope that developing countries would follow in due course. Put simply, the focus was on emissions in those countries where emissions were not growing very much – rather than on the countries where they were growing very rapidly.

Worse still, whilst this unprecedented expansion of coal burning was going on, and with emissions marching ever upwards, many political leaders seemed to think that the problem could be best addressed by building wind farms and putting solar panels on the roofs of houses – and putting insulation in the roofs of houses too. It was as if either all the carbon-intensive, coal-based goods from China didn't matter, or they could be left for another day. Unilateral action was the order of the day, as if global warming was a national event.

It would be hard to make this up, but sadly it is true – appallingly true. The Kyoto Protocol cemented this approach, and the green non-governmental organizations (NGOs) and green parties that sprang up in Europe jumped at the chance to build their preferred decentralized societies. Europe would 'lead by example', and Connie Hedegaard, the European Commissioner for Climate Action, could claim that the EU has been 'in the vanguard of international action to combat climate change', and that its targets 'have given Europe a head start in the race to build the low carbon and energy efficient global economy'.[3] If only it were true. Indeed, as we shall see, EU policies have achieved at best little. They have probably made things worse in fact – and at great cost. Europeans reinforced all this by concentrating on a short-term target of moving towards 20% renewable energy by 2020 to meet their 20% emissions target. The economic illiteracy is astonishing: for a long-term problem the Europeans chose a short-term target, and seem to have convinced themselves that the target and renewables and energy efficiency should all add up to the same magic number. Quite why everything should add up to '20', and quite what all the current wind and solar technologies would add, when set against the great coal-burning expansion, remains a mystery – and one this book seeks to explain.

No doubt our historian in 2100 will conflate the European attempt to strut the world stage on climate change with its mistakes on currency union and the Eurozone crisis. Both share the same hubristic optimism: that announcing a policy solves a problem. Not everyone was fooled by Kyoto. The US never ratified the Protocol – for the very good reason that China would not play ball. What exactly is the point of reducing emissions in the US (and Europe) if it encourages energy-intensive industry to move to China, where the pollution will be even worse? Russia joined only because it would get paid to do very little, and a number of other countries joined but did not really try – Canada, for example.

So it is not surprising that under Kyoto nothing has been achieved in terms of global emissions. They keep on going up relentlessly. This will no doubt puzzle our historian. How could so much effort by expended to so little effect? What will be even more puzzling is why, despite the lack of any serious impact, world leaders kept their foot on the Kyoto pedal, as if a few more pushes would somehow overcome all the problems.

This reached a farcical climax at the Copenhagen conference in late 2010. Imagine the scene. The runway is full of jets from all the campaigners and world leaders. Europe offers a 30% cut in its emissions if others follow. A huge media circus envelops the city, with proclamations and statements from green NGOs and interested parties. And what happens? At the last moment, the US and China cut a deal outside the framework of the talks, promise to try to limit warming to 2°C, and promptly leave. Thus was born the Copenhagen Accord – to sink almost unnoticed in the months to come.

Our historian might ask a very rational question: faced with such an obvious setback, did the players go back to the drawing board, ask whether the whole approach was flawed, and reconsider how to tackle the problem? Not a bit of it. Connie Hedegaard concluded that Copenhagen 'tells me that such leadership by example works'.[4] The conclusion the Europeans drew – along with the green NGOs and green parties – was that they just had to try harder. It could not be that the approach was wrong. More of the same was what they thought was needed.

Another year went by; another 2–3 ppm emissions were added; and the expiry date of the first period of the Kyoto architecture at the end of 2012 loomed. Now all eyes were on the climate conference in Durban at the end of 2011. If Copenhagen went badly, Durban was a disaster. At the end of another circus of green NGOs, media and publicity-hungry politicians, what was 'agreed' was that the parties would try to agree by 2015 what they might do after 2020. This really would be hard to make up! They could not even agree what legal form that which they might eventually agree might take. And this was a triumph!

We have come to an impasse. On the current plan, nothing much is going to happen for another decade, 30 years after the starting gun was fired. By 2020, short of a (possible) major setback, China will have doubled its economic size yet again, as will India. 400–600 new coal-fired power stations will be added to global electricity generation. Emissions will almost certainly pass 400 ppm. If the mainstream scientists are right, there will be no chance of limiting climate change to 2°C, and we will probably be committed to much more.

No doubt by then there will be lots more wind farms in Europe, but our historian will be able to observe what has really been going on in Europe. Faced with what its leaders think of as a major threat, the Germans decided to enhance their contribution to tackling global warming by instantly closing eight nuclear reactors in 2011 (following the Fukushima nuclear power plant incident in Japan), presumably on the grounds that a tsunami might reach Munich. This low-carbon generation had of course to be replaced, and Germany fast-forwarded its development of more fossil fuel electricity generation – including two massive new lignite coal power stations. Lignite-based electricity generation is about as dirty as you can get.

The Germans had their reasons, and they were not alone in their ambiguous response to global warming. The British decided to push on with one of the most expensive ways of generating low-carbon electricity known to man – intermittent offshore wind – and, copying the Germans, found an even more expensive option: rooftop solar for its northern climate. Our historian will be understandably puzzled: how could such expensive options be chosen first? Moreover, what is the question to which offshore wind and rooftop solar are supposed to be the answer? It can't be global climate change – wind farms in the North Sea will make no difference to climate change. They will not even make much difference to Britain's carbon footprint. It's a question of basic arithmetic: even with the larger wind turbines, hundreds are needed to match a single gas, coal or nuclear power station. Indeed, they may actually exacerbate global emissions as they drive up energy prices and thereby encourage further deindustrialization at home, relying on imports of carbon-intensive goods from abroad.

Thus despite innumerable conferences, summits, proclamations, agreements and policy interventions, so far nothing much has been achieved, and indeed some of the interventions may have made things worse. From a climate change perspective, it has all been pretty disastrous. By 2020, three decades will have been largely wasted. That is the bad news, but it need not be like this going forward. Climate change is not an insoluble problem, and there clearly is some willingness amongst the wider populations to tackle it. Resources for tackling climate change are scarce, but at least there are some, and the task is to deploy them in the best possible way.

To get from the negative to the positive, it is necessary to focus relentlessly on the nature of the underlying climate change problem: why it matters, what causes climate change, and who is responsible. Part One of this book is about sorting out these fundamental building blocks. Amazingly, for the most part, the debates on climate change have studiously ignored them. Instead of homing in on coal, China, economic growth and the underlying population growth, the emphasis has been on the *production* of carbon emissions in Europe. What matters – the carbon footprint – has been largely ignored. Whilst Europe has been deindustrializing its own production, it has not decarbonized its consumption. Indeed, once imports of carbon-intensive goods from countries like China are taken into account, the reality is that Europe's carbon consumption has been going up. In Britain's case, as we shall see, focusing on carbon consumption reveals the true scale of the deception: whilst carbon production fell by 15% from 1990–2005, carbon consumption went up by 19%. Rather than boast of their achievements, our political leaders should hang their heads in shame. A similar disastrous story is writ large across Europe. Thus Connie Hedegaard claims that great credit is due because the fall in carbon emissions in Europe between 1990 and 2009 has been 16% whilst economic growth was 40%, and that this shows the world that 'it is perfectly possible to cut emissions whilst growing the economy'.[5]

Yet it is all smoke and mirrors. No mention of the consumer-driven nature of that growth; no mention of all the carbon imports which fed that consumption. But to admit that Europe's carbon consumption was going ever upwards would spoil her story and of course paint a very different picture.

Getting a grip on the true nature of the problem is the first step to solving it. The next is to understand why almost everything the developed countries (and especially Europe) have been doing has not, and will not, crack the problem. *Current* renewables (and it is crucial to recognize that there is a very important distinction between current and *future* renewables) and current energy efficiency policies (based on *current* technologies) will not close the gap. Wind farms and energy efficiency measures have a part to play, but alone they are wholly inadequate for the task. A North Sea coated with windmills will make little difference at enormous cost, and energy efficiency measures will not stop energy demand going up. Part Two explains why this arithmetic does not – and cannot – add up.

It remains a puzzle as to why these policies have nevertheless been pursued with such zeal by political leaders and green NGOs. How could they be so misguided? The answer, as we shall see, turns on a highly questionable assumption in the climate change policy debates. Advocates of *current* renewables frequently claim that although they are currently expensive, they are going to become cheaper than fossil fuels since the price of oil and gas is going to go up because we are running out of them. It is claimed that fossil fuel production will peak, or has already peaked, and hence we have no option but to go for *current* renewables.

Sadly, the contrary turns out to be the case. Far from running out of fossil fuels, we have more than enough to fry the planet, and in Part Two it is explained why, for at least policy purposes, we should probably assume that the supply of gas is infinite, and why oil and gas are more easily substitutable than the peak oil brigade imagines. It is just wishful thinking that we will be forced to decarbonize because we run out of fossil fuels and, more immediately, somewhat irresponsible to assume that oil and gas prices will go ever upwards.

The full context of the policy failures have to be understood in order to make progress. Good people tried to make it work. They tried, but for the most part they clearly failed. At the heart of this is the Kyoto Protocol and the efforts to build a credible, binding legal agreement around it. The reality is that they could not have succeeded – and the tragedy is that they didn't realize this, and as a result wasted years of effort and political capital. There is unlikely to be a comprehensive global agreement that sticks any time soon, and in particular not soon enough to make a significant difference to climate change. Why? As we shall see, it just isn't possible to craft an international agreement that is binding, credible and enforceable on production targets.

Understanding why climate change matters, what causes it, who is responsible and why so little progress has been made is a necessary, and cathartic, exercise because it reveals the underlying raw nature of the awesome problems we face in the early twenty-first century and whether there is a chance of providing a happier ending for our future historian to write about. We cannot go on like this, and Part Three sets out how climate change can be cracked.

This requires focusing on what we are trying to do, on doing it at least cost, and using human ingenuity to build a more sustainable economy. Politicians like to promise solutions without pain – and then go on to impose much more pain than they need to precisely because they avoid explicit prices and costs. The problem of carbon emissions will not get addressed unless carbon is integrated directly into economies and economic decision-making. This rather obvious point has some very powerful implications which will be developed in Part Three. It means that we must pay for the carbon emissions we cause, and hence for our carbon consumption and not just our carbon production. That means not just domestic carbon prices, but also the much more controversial question of the pricing of carbon embedded in what we import. It is easy to say that having a carbon price is a necessary condition for tackling climate change. It is much more difficult to live with the full implications.

Integrating carbon into the economy enables choices to be made – primarily by markets – about how to generate electricity (which is increasingly the final fuel of choice) and to power transport. That means making choices not only between fossil fuels and renewables, but also between different types of fossil fuels. Getting out of coal is an absolute and immediate priority, and Part Three explains how gas may play an important transitional role, and what shale gas means for climate change and climate change policies. For while all eyes have been on the promise of renewables, a revolution in fossil fuel technologies has taken place. Huge quantities of hitherto uneconomic gas supplies have become available, transforming not just fossil fuel markets, but geopolitics too. This changes the game. It cannot be reversed, and any serious energy and climate change policy has to come to terms with the enormity of what has happened, rather than either ignoring it or wishing it would go away.

Integrating carbon into the economy and switching from coal to gas will help, but they are means to an end: a low-carbon economy. The final part of the argument lies with new technologies, and Part Three showcases some of the plethora of possible advances. The laboratories are full of new developments, including batteries and storage, smart technologies, and new-generation technologies. This is where the distinction between *current* and *future* renewables comes in. If current renewables cannot crack the problem, it does not follow that future renewables will not. On the contrary, these developments offer great possibilities. But technological progress will not happen by accident, although carbon pricing will help. Rather it will take money – and in the current economic difficulties, money comes from hard-pressed taxpayers and consumers. Money spent on one thing (like current renewables) is money not spent on something else (like future renewables). There are choices. That is what economics is all about – allocating scarce resources. It's a reality that too many politicians (and many green NGOs and green parties) ignore. Economic illiteracy is at the heart of the climate change problem, and although it can be solved, it won't be if we go on wasting so much money to so little effect.

PART ONE

Why should we worry about climate change?

CHAPTER 1

How serious is climate change?

Why don't we take climate change as seriously as scientists tell us we should? Climate change ought to be something that we can easily grasp. It lends itself to disaster movies,[1] to iconic photographs of polar bears on small lumps of ice in melted seas, and to glaciers collapsing into the sea. Droughts, floods, hurricanes and heatwaves capture the imagination.

It is not as if the scientific evidence is getting weaker. On the contrary, the science continues to progress – as do the emissions – and the forecasts have not got any better. The latest annual update from the International Energy Agency (IEA) – the 2011 'World Energy Outlook' – boldly stated:

In the New Policies Scenario [in which recent government commitments are assumed to be implemented in a cautious way], the world is on a trajectory that results in a level of emissions consistent with long-term average temperature increase of more than 3.5 degrees C. Without these new policies, we are on an even more dangerous track for a temperature increase of 6 degrees C or more.[2]

There is a paradox here: the public mood gets more indifferent or even sceptical about taking the necessary action, whilst at the same time the science – and its media face – becomes more apocalyptic. Compared with the last decade, fewer people in many of the major developed countries now report a willingness to pay much to mitigate emissions, even amongst those who accept the science. To this public scepticism has been added an element of 'climate change fatigue'. They read the stories, see the pictures and the movies, and yet nothing dreadful seems to happen, and nobody else seems to be doing much about it. Crazy as it seems to many main-stream scientists, many still wonder whether climate change really matters, and even if it does, whether it matters very much. That makes it much more difficult to crack the problem, since it is the broader public that must pay the bills for mitigation, not the scientists.

Instead of blithely dismissing this as the malign influence of mavericks and climate change 'deniers', it is important to understand precisely why so many have their doubts, and why it is so hard to get any action. Only then can we take on the business of explaining just why climate change is so serious, what we do and do not know about the underlying physical proc-esses, and what the likely impact of various degrees of warming might be.

Why the public has doubts

Some climate scientists and some economists are partly to blame for this scepticism and fatigue. It is corrosive: politicians react to public opinion, and as they see the voters' interests waning, they follow suit. Scepticism comes in three forms: about the science, about the economics, and about the willingness of others to do much about the problem. It shows up in opinion poll surveys. Across Europe by the end of 2011, electorates' warmth towards green parties appeared to be on the wane – even in Germany, where the Green Party's share of the vote fell back from the post-Fukushima highs of around 25% to below 15%.[3] Indeed, in Germany the Pirate Party achieved almost 9% in Berlin in regional elections in late 2011, and was almost level-pegging with the Greens in state elections in early 2012. It is difficult to interpret what these swings mean, but the volatility does lend credence to the view that many vote green as a protest rather than as a deep commitment to what these parties advocate. Elsewhere in Europe, the far-right political parties, which tend not to worry too much about the climate, have been scoring high ratings.

Climate change has slipped far down the list of voter priorities. In the US, the Center for Climate and Energy Solutions (formerly the Pew Center) reported a marked decline between 2006 and 2009 in the number of people who see solid evidence of global warming.[4] Others have provided more positive results.[5] Yet overall, given the advances on the science front, far more positive results might have been expected.

Of the three forms, scepticism about the science is the most serious. After two decades of further research since the UNFCCC, we now know a great deal more about the theory and the empirical evidence. Scientists have made great strides in the physics, in looking at ice cores, ocean currents, sea levels and changes in the Arctic and Antarctic.

Yet the public, ignorant of the complexities of the science, have taken a good hard look at the messenger and have questioned the basis of this developing science. When they are told that 'scientists say . . .' it is not unreasonable to ask: what are their interests? Have they got 'skin in the game'? The grants, the money, and indeed the career prospects and the status go to those who conform to the paradigm. Those following the scientific consensus tend to control the key journals; they are the heads of the departments, and they are members of the grant-giving bodies.

As has occurred all too often in the history of science, outsiders holding sceptical views face an uphill struggle to get themselves taken seriously. Aided and abetted by the green groups and lobbyists for particular technologies like wind, the way to undermine those who challenge the conventional wisdom is to question their motives, discredit the individuals, create a 'them and us' division, and never to question the mote in their own eyes. Many greens cut their teeth on the all-out assault on nuclear power. Sceptics become deniers. Nigel Lawson found it hard to get his book on climate change published, even though he accepted that climate change was likely to occur.[6] Bjørn Lomborg became a pariah, again despite accepting that warming would probably happen.[7] These writers found that even questioning the relative importance of climate change was unacceptable not only to activists, but also (and more worryingly) to sections of the mainstream media. All this is frighteningly akin to the treatment of minorities in the wider social and political sphere. Not surprisingly, the public smell a rat.

Climate scientists, often passionate in their concern for the future, and watching emissions going ever up, are all too human. They want to 'do something'. They take to the airwaves and the televisions screens, and the boundary between science and NGO-style campaigning begins to break down. The seminars at environment and climate change institutes and departments invite in the campaigners, blurring the boundaries between impartial academic research and partial and committed interests.

The University of East Anglia (UEA) 'Climategate' affair illustrates what can go wrong. Hostile critics got hold of emails from UEA, which revealed a number of worrying features.[8] Accusations were made about data manipulation, and about an approach to research that appeared to make sure that the evidence lines up with the preferred conclusions. Opponents claimed to be deliberately undermined. The result has been inquiries and reviews, none of which has concluded that much more than the occasional lapse occurred. Yet it is hard to argue that the treatment would have been the same had the emails come from someone questioning immigration, for example, or considered to be pro-nuclear.

The media's habit of ascribing extreme weather events as evidence of climate change, often encouraged by interested scientists, has not helped. Whilst scientists tend to be careful not to claim explicitly that hurricanes like Katrina in 2005 or the European heatwave in 2003 were *caused* by climate change, they use these examples to illustrate what they claim will happen.[9] There is an implicit linkage – and too often exceptional cold weather is met with a deafening media silence. Again the public notices.

Scientists cling to the defence that 'peer review' ensures that they keep to the straight and narrow. This might convince some within the academic community, but it does not ring quite so obviously true to the wider public. Taken seriously, peer review is an open process by which new research is tested by other scientists. The approach should be sceptical – trying to find reasons why claimed results may not stack up. Yet when it comes to climate change, it has not always been rigorously adhered to.

Even the most important body on climate change, the Intergovernmental Panel on Climate Change (IPCC), has had its lapses. The IPCC is an unprecedented endeavour. The idea is to bring together the best scientists in the world to investigate the science of climate change and to provide authoritative conclusions for governments and the public. From the outset it was designed to support advocacy of urgent action. It is both heroic and questionable. It is heroic in the sense that it tries to corral disparate research groups. Academics are notably anarchistic. Groups in different universities compete in the world of ideas and evidence. It is not in the nature of the beast to be brought together under one global umbrella. But the IPCC results depend upon who does the science, and it is a voluntary activity, typically unpaid and often without grants. It is also immensely bureaucratic. Those who most enthusiastically join in are likely to be those who believe most strongly in the outcome. So there is a problem. Those on the inside tend to do the peer review. The result is to reinforce 'group-think'. The IPCC approach has sadly been characterized by some of these paradigm effects. Claims about the melting of Himalayan glaciers turned out to be not only exaggerated, but also inadequately checked.[10] A more recent example is the IPCC's 2012 report on renewable energy sources and climate change mitigation.[11] It is a struggle to find anything other than positive arguments for renewables, as opposed to objective analysis, which would include more on costs, limitations of land areas, and a serious consideration of environmental impacts. Fritz Vahrenholt gives an alarming account of what happened when he was asked to review the renewables report in 2010:

> I noted many errors and in the end someone from Greenpeace edited a main part of the summary of the report. A Greenpeace scenario, claiming that in 2050 we can produce 80% of our energy with renewables, was presented as one of the main considerations.

He goes on with his devastating critique: 'I found out that one third of the core writing team of the summary for policymakers in 2007 had connections with Greenpeace and WWF'. Whilst correctly pointing out that this does not prove that the report is false, he asks this telling question: 'Suppose that people find out that the IPCC summary reports were written by people with connections with Exxon or Shell – would that be acceptable?'[12]

But at least the IPCC's work is supposed to be peer-reviewed. One might conclude that anything not peer-reviewed in the climate change literature should be discarded – not to be taken seriously because it has not endured trial-by-peers. But this is not so, which brings us to the second example, and the second area where scepticism has emerged – on the economics front.

Nicholas Stern's *The Economics of Climate Change* (hereafter referred to as the Stern Review), published in 2007, is one of the most referenced studies.[13] Every major European politician has cited it, and it is referenced at the back of numerous books and articles on climate change. Yet it was not peer-reviewed. Although led by an eminent economist, it is the work of a government-sponsored and government-resourced endeavour, undertaken in a very political context (Prime Minister Blair's positioning at the G8 Gleneagles summit, and Chancellor Brown's desire to capture the issue). As William Nordhaus put it, 'the Stern Review should be read primarily as a document that is political in nature and has advocacy as its purpose.'[14]

None of these considerations is a criticism of the content of the report itself (although, as we will see, there are many reasons to question both its analysis and its conclusions). Rather, the 'peer review' standard turns out to be regarded by mainstream scientists as unnecessary when it suits them. Indeed, they widely quote the Review.

It is easy to see why both politicians and scientists got so excited by the Stern Review – and the activists too. It concluded rather precisely:

An estimate of resource costs suggests that the annual cost of cutting total GHG [greenhouse gases] to about three quarters of current levels by 2050, consistent with 550 ppm CO_2e[15] stabilization level, will be in the range −1.0 to +3.5% of GDP, with an average estimate of approximately 1%.[16]

When compared with the costs of the credit crunch and the Great Recession that followed, it is a very manageable number. But imagine if the Review had concluded that the economics of climate change pointed to a very high cost of mitigation. The view from both the academic community and especially the green NGOs would surely have been that it could not be taken seriously because it was not peer-reviewed.[17]

Politicians jumped on its conclusion. The 'greening' of the economy through climate change policy has been variously claimed to be a way out of recession, the path to a 'new industrial revolution', and capable of lowering future energy prices. Listening to the President of the European Commission, José Manuel Barroso, you could easily be led to believe that the challenge of climate change is actually a good thing regardless of the climate; that climate change mitigation will lead to future prosperity. Here he is in New York in the run-up to the 2009 Copenhagen climate conference:

> In fact, the economic case is just as strong [as the scientific]. Acting against climate change is of course a moral imperative to current and future generations. But it is above all an immense economic opportunity. The cost argument is getting clearer. Tackling climate change will be expensive ... But it's about more than minimizing the pain; it's about surfing the next wave of economic development. Take Europe's climate and energy package, for example, which we agreed in 2008: cutting emissions by at least 20% below 1990 levels by 2020 and doubling the share of renewable energy to 20% within the same time-frame. We think this will generate some €90 bns ($130 bns) of additional investment in renewables, and some 700,000 new jobs in this sector, as well as reducing our oil and gas import bill by around €45 billion ($70 billion) a year by 2020.[18]

Such claims stretch economic and political credibility, especially when they appear to jar with customers' actual bills. Politicians tell us that the solution to the economic crisis is 'green growth', and even that decarbonization will reduce energy bills by 2020. The mantra about the sunny uplands of decarbonization just keeps on getting trotted out. It's hard to take seriously – that the world's carbon-based economy can be decarbonized in a few decades without economic pain; that we will all be better off. Even more surprising is that apparently intelligent people actually seem to believe it.

The third source of scepticism derives from the fact that others appear to be doing very little. Many who may accept the scientific consensus, and who may even agree with Stern about the economics, nevertheless remain sceptical about the actions of others, notably the US and China. This is understandable given what has happened so far. The Durban conference at the end of 2011 confirmed this public suspicion: nothing of substance is scheduled to happen before 2020.

Scepticism is corrosive: if the public doubts the science, questions the economics, and despairs about the inaction of others, then the climate change battle will be lost – and indeed it increasingly looks as if it will. To recover the initiative from this sorry state of affairs, we need to sort out what we do know from what is hypothesis, so that the uncertainties can be understood, together with their implications for policy. It turns out to be anything but straightforward.

The science of climate change

The first step in the science is the easy bit: laboratory experiments can be conducted with stylized atmospheres to work out what happens when the concentrations of different gases are increased. This is 'old science': in the nineteenth century, the 'greenhouse effect' was identified by looking at the penetration of the sun's rays and reflection back through a mixture of gases, of which carbon dioxide (CO_2) is a very small but key component, along with a number of other greenhouse gases, especially methane, and water vapour. The greenhouse effect was suggested long ago by the research of Joseph Fourier and then John Tyndall in 1859.[19]

That is step one, and no one seriously disputes this finding. CO_2 (and other greenhouse gases) is crucial to keeping our planet habitable. Without it the earth would be largely frozen, but there are good reasons for thinking, all else being equal, that an increase in carbon in the atmosphere will increase the greenhouse effect, trapping heat at the earth's surface, and reducing the amount of heat that radiates back. A little carbon is a good thing: too much is a bad thing – potentially very bad.

But it is a much more contentious step to suggest that this is what actually happens when the laboratory gas is swapped for the atmosphere, and all else is not necessarily equal. This is when all sorts of complexities are brought into play: as carbon emissions rise, so do feedbacks, and vice-versa. Plants may take up more carbon through photosynthesis; the resulting heat may lead to more clouds and rainfall. Water vapour concentrations change. Heat may change the oceans' currents. More thawing of snow and ice in summer may lead to a reduction in the surface areas of the planet that are covered in reflective white, which bounces back heat (the so-called albedo effect). The reduction of the sea ice allows the sea to absorb more heat, creating a double whammy – darker surfaces absorb and retain more heat. This is one reason why the Arctic is warming so fast. (Conversely, when carbon concentrations fall, water vapour levels fall, leading to more cooling. This process may have helped to trigger the ice ages, as Svante Arrhenius set out in 1896.[20])

The change in the composition of the atmosphere is not just an increase in CO_2. Again, all other things are not equal. All sorts of other pollutants are being added over time, not least from coal-burn. Higher levels of sulphur dioxide might also have an impact on temperatures (as well as causing other environmental damage). The atmosphere comprises a delicate balance of its overwhelming components of oxygen and nitrogen, and other key gases. It is anything but simple.

Once we move from theory to the world of predictions, the question of data and evidence comes to the fore. The obvious place to start to forecast the future is to see what can be discovered about the past – when carbon concentrations and the global climate both varied (as indeed they always have). In particular, did the recent (in geological terms) ice ages coincide with less carbon, and the warmer interludes (one of which we probably live in now) with more carbon? Which way did the causation go – from carbon to temperature, or the other way round?

Ice cores provide a window on our past climates and they go back 800,000 years right through the geologically recent ice ages, and we can hypothesize from the data what the temperature and atmospheric changes might have been like at different times. They tell us that the ice ages and the concentrations of carbon are indeed correlated. And we can try to go back even further in geological time by looking at mud cores in the oceans, and consider all sorts of other clues to our climate past such as pollen and peat bogs.[21] It's a great detective story – assembling the clues and trying to make sense of the disparate bits of evidence.

Together these give us *deep* climate history – geological history. Correlation and causation are not, however, the same thing: temperature change could have caused changes in atmospheric gases rather than the other way around. Then there is the question of what caused the initial changes in either temperature or carbon concentrations (or both) when there were no humans around. Carbon concentrations have varied a lot, as have climates, and therefore there must have been other reasons for the atmosphere to change. Variations in solar radiation and the tilt and orbit of the earth play a key part. So too do volcanoes. Plants have transformed our atmosphere, adding oxygen and subtracting carbon. Climate change obviously does not have to be man-made.

Once we recognize the possibility of other, non-human, causes, there follows a search for them, and it is an easy step for sceptics to suggest that much of *current* climate change might not be man-made either. A host of hypotheses arise, and since we know that there has been non-man-made climate change in the past, they cannot be rejected out of hand. It can work both ways – raising or lowering temperatures. Thus if the temperature predictions of climate scientists for the last decade have not been realized, it could be because of other offsetting cooling factors.[22]

Because other factors can cause climate change, it does not follow that increased emissions generated by man will not have caused it. The trouble with the man-made bit is that the data covers such a short period – from the Industrial Revolution. It is a mere 200 years, during which time carbon emissions have increased substantially. Before that climate changed a lot too, especially on a regional basis. It always has.

The data from the Industrial Revolution onwards on climate and temperature is itself not clear-cut. The quality of temperature measurements in Europe improved and records were kept, but European climate change is not the same as *global* climate change. What happens at the global level is what matters, and for this we need records of worldwide temperature changes. Here the picture is much more patchy.

Temperature readings need to be clear of interference. They cannot, for example, be from weather stations in areas where there is urbanization going on, since this will change the background conditions. These relatively short-term records have to be painstakingly put together, and the creation of a global average temperature is a statistical construct from these numerous – and inevitably imperfect – data locations.

So we have a theory tested in the laboratory; we have atmospheric models; we have geological data; and we have recent detailed temperature records. Climate science has been a gradual process of unpeeling a world of enormous complexity. It is not simply a matter of having a theory, testing it, and confirming a 'truth'. Rather it is a process of establishing a conjecture, and peeling back the detail. It is this messy reality of the science that is the basis of policy, not some near-certainty that will become certain if we just do a bit more research (although that of course would help).

Bringing all this detail together in integrated climate models is much more complex than doing a laboratory experiment, and it is to be expected that knowledge will be partial, and that different models will produce different predictions. To expect something as complex as the atmosphere and its relationship with climate to result in computer simulations looking a century ahead which have common results is not just misguided, but is actually to misunderstand the nature of science and the scientific process. These models are unlikely ever to agree, in the sense of establishing 'facts' that prove the claim that global warming is the result of increased emissions. Science deals with uncertainty, not certainty – and good scientists are at pains to point out the limitations of their tentative knowledge.

What the models show is a series of predictions for temperature changes for different concentrations of carbon (strictly greenhouse gases, but we continue here with our shorthand of referring to carbon). Not surprisingly, the outputs – the temperature predictions – depend on the model inputs, and the result is a temperature range of anything from 1.1 to 6.4°C by 2100 compared with the UNFCCC baseline of 1990.[23] Precise predictions not only lack scientific credibility, but also stretch public credulity. This range is anything but precise, with an enormous spread from a manageable increase to catastrophe. The sceptics and the catastrophists focus on the outlying extremes – the former claiming that climate change doesn't matter, and the latter that Armageddon looms.

The best we can have is a tentative or conjectural knowledge.[24] The general conjecture that increased concentrations of specific gases in the atmosphere are causing global warming is one we can use for policy purposes. If we ask not what is the predicted temperature increase by 2100, but rather, in looking into the scenarios in detail, whether there are any good reasons for thinking that, on a business-as-usual basis, it will be less than, say, 3°C, then we can work on whether action is worth taking now to avoid the risk of at least 3°C. If this condition is met (as the IEA thinks it will be), then whether it is 3°C or 6°C is not the issue: 3°C is, as we shall shortly see, bad enough to merit action, which would also address 6°C. On the other hand, if 1°C looks likely, there are probably more important global problems that we should concentrate on. A great deal of research (and emitting) has gone on since the last IPCC projections in 2007. The bad news is that, for all the caveats and uncertainties noted above, 1° looks unlikely, and that 3°C+ is now supported by a considerable amount of evidence. There is then a non-negligible possibility that we are in for significant warming. That is the best we have to go on.

How bad might it get?

Assuming that increased man-made emissions are altering the climate, the question the more agnostic public want answered is: does it matter? By this they mean: are we sure that climate change is always and everywhere a bad thing, and if it is, is it a *very* bad thing? Many scientists seem to consider the answer obvious, and indeed seem to regard even those who raise the question as guilty of heresy. Yet the question has a lot of merit, and it is not, in fact, easy to answer.

One way of tackling this is to start with a small change in climate – a gradual increase in global temperatures. Given that most of the world's wealth lies in temperate zones (notably in the US and Europe), and given that winters can be cold, and require a lot of energy (heat) to be expended to mitigate the impact of the cold, a small increase in temperature might well be economically advantageous, not least in cutting our energy bills.

This observation has historical resonance. The climate has always been changing, and within human history there have been substantial swings in global and regional temperatures. Indeed, much of human history has been shaped by these changes. The warmer climate that prevailed in Europe during the Roman period helps explain how agriculture could sustain the northern corners of their empire. It was much warmer when the Vikings colonized Greenland. It had to be; otherwise they would not have been able to cultivate the land, and they would not have made it to the American continent.[25] After around 1300 it got a lot colder. The Vikings in Greenland were eventually wiped out, Europe suffered from famine, and even the River Thames froze in winter.[26] It was only in the nineteenth century that the European 'Little Ice Age' came to an end – and not necessarily because of the early years of the Industrial Revolution.[27]

Longer growing seasons and milder winters mean that the constraints on economic activities in the cold places (and the costs) are somewhat alleviated. Experiences in the tropics might of course be very different, although not obviously so. A small rise in temperature does not necessarily mean less rainfall, or longer and more intense droughts. On the contrary, it could add to agricultural output. More carbon in the atmosphere might actually be good for plant growth. For the world's largest economies – and its most energy- and carbon-intensive ones – climate change is initially likely to be good news.[28] A gradual increase in global temperatures might be accompanied by a faster increase in temperature in the Arctic due to the albedo effect, as noted earlier. The results here would be dramatic – much more so than almost anywhere else on the planet.

Is a gradual increase in average global temperatures and a faster increase in the Arctic a bad thing? Again the answer is far from clear-cut. It will be bad news for the polar bears and other specialist species in the Arctic – although in biodiversity terms, the opening-up of the ice may benefit some species now limited by the ice cap (very cold environments are not known for the richness of their biodiversity). Changes in climate have the greatest impact on marginal species, which tend also to be specialist.[29] Similar effects were felt at the end of the last ice age – which was bad news for woolly mammoths. Climate change is disruptive, but its very disruptiveness is what spurs evolution, and it was climate change that enabled humans to spread across the planet as the ice retreated.

The ice is a major barrier to accessing significant raw materials – oil, gas, coal and minerals. Around Greenland, the opening-up of Arctic resources is gathering pace, and the local population is on the cusp of a revolution in its economic circumstances. Whilst others might romanticize the primitive life, the Arctic peoples have had to live in such harsh environments, and it is not clear that they see the new wealth as entirely negative. The famous Northwest Passage between the north Atlantic and the Pacific (and the Northeast Passage too) becomes a distinct practical option for shipping. Mineral resources become accessible on an enormous scale. The Arctic may contain perhaps a quarter of the earth's undiscovered conventional oil,[30] and the isolated examples of mineral extraction – like coal mining in Svalbard – may be replaced by large-scale industrial extraction. The Greenlanders – all 60,000 – may become extremely rich, and hence the scramble to establish who owns the Arctic seas. From an economic perspective, the five Arctic nations now regard them as valuable potential resources rather than a white, cold desert. In addition to the Greenlanders, Canada is a rising, resource-rich country, Alaska has already been opened up to oil, and Russia's vast Siberian assets are becoming more easily accessible.

There are downsides here too: melting permafrost is not easy to cope with, and there will be lots of localized impacts. But on balance it is hardly surprising that at the Durban conference in late 2011 Canada indicated a preference for withdrawing from the Kyoto framework, and Russia was at best half-hearted. No Arctic nation (unless Denmark speaks for Greenland) is very actively campaigning for major mitigation measures: all (even Denmark) are actively supporting oil and gas developments.

If the case against a bit of global warming is fairly weak, what about quite a bit of global warming – say 2°C? This is a much more interesting and difficult question, and has captured the imaginations of many. Given that there is little hope of getting back to pre-industrial levels, could carbon be stabilized at around 450 ppm? And if it cannot, how bad might the consequences be?

There is nothing magical about 400 ppm (or 450 ppm CO_2e) or 2°C (or the fact that 450 CO_2e ppm has been equated to a temperature increase of roughly 2°C). But it does have one merit: some have argued that it might be just about achievable (although it is now unlikely), and it is for this reason that 2°C has become the de facto target (and indeed, after Copenhagen, a more general aspiration).

What would happen at 2°C warming on average? Bearing in mind the uncertainty inherent in the science and in the climate models, the first important thing to say is that we have only a very general idea – and one that is fragmented at that – even though we have been there a number of times in human history.

We could have an ice-free Arctic summer by 2050, or perhaps even by 2030. Sea levels would rise a bit – not the two metres often cited in the media, but several centimetres. The Greenland ice sheet would take quite a long time to melt. A small rise would cause coastal flooding to some areas – many of these being relatively economically poor, like

Bangladesh. Coastlines can be protected: around one third of the Netherlands is currently below sea level, but is kept dry by sea walls and pumps. London might need a new, larger barrage, but the costs would be manageable. Similarly, there would be problems on the eastern seaboard of the US, but again the adaptation measures are not beyond its economic capacity.

When we break down the effects in these ways, what at first looks like the alarmists' Armageddon – and is often presented in this way – turns out to be a set of economic costs (against which benefits have to be weighed). What this sort of consideration leaves out is the more dramatic effects: floods, droughts, storms and plagues.

The trouble with effects of this type is that we really are quite uncertain about how they will play out. Weather, like climate, is immensely complex. That complexity comes from the interplay of wind, ocean currents, temperature, cloud and sunshine, mountain ranges and so on. Think of some of the implications of higher temperatures. It could be a wetter world, with more water vapour and rainfall, whilst some areas could be drier. It could be more variable. Deserts might get greener – parts of what is now the Sahara grew grain for the Romans, in a period when it might have been up to 2°C warmer.[31] There are cave paintings in the middle of what is now the Sahara. On the other hand, the ocean currents may change their flows and directions. There has been much speculation about the slowing of the North Atlantic Current because there will be more cold (fresher) water from ice melting in the Arctic. That would make northern Europe colder, yet the oceans on average would be warmer. But again this is uncertain, and in any event, a cooler western Europe might offset some of the warming impact.

The case for urgent action to deal with the possibility of 2°C varies according to the different countries' locations. The costs would be very unevenly distributed. But what worries scientists – and indeed what should worry everyone – is what happens if the temperature increases go *above* 2°C, and in particular if they keep going up. This would bring us into temperatures not experienced on earth for the last 800,000 years. And this is what the models, for all their weaknesses and uncertainties, imply that we may be in for.

Intuitively, the idea of a runaway increase in temperatures does not seem very plausible. Climate varies a lot anyway. There are numerous factors at play and, as noted above, climate change pre-dates man. In principle, there are good reasons to believe that whilst it is always changing, there may be equilibrating forces at work, bringing variations back to a mean. It was warmer before 1300 in northern Europe, and then it was colder, and then it got warmer. More recently, the downturn in temperatures after the Second World War led a lot of scientists to demand action to prevent another ice age. (Some of these scientists are the same ones urging action to reduce global warming now.) Many people think in terms of natural cycles, and evidence from things like the earth's tilt and sunspot activity lends a credible scientific explanation to such cycles. The ice ages themselves might have had such causes.

But there are more worrying dimensions to take into account. The balance of evidence suggests that rising temperatures are not being triggered by a natural process that might be reversed, but primarily by a man-made one which can probably be reversed only by deliberate human actions. The carbon locked in the earth's crust is being transferred to the atmosphere. In time, photosynthesis might take the carbon back. More water vapour in the atmosphere caused by higher temperatures might even produce more clouds, causing a drop in temperatures, and thereby creating 'snowball earth'.[32] Natural processes like photosynthesis will take a very long (geological) time – too long for the human race, a species that has not been around for even a million years.

And in the meantime, water vapour, the ice melting and the albedo effect might create positive feedbacks – most alarmingly the release of methane from the tundra. If these were to occur – if we were to have rapid climate change – then there would not be time to adjust and the economic impacts might be dire. The rise in sea level if the Greenland ice sheet melts much faster then makes a lot of difference, and the shift upwards of, say, 4°C renders many current marginal areas anything but marginal. It would take an awful lot of technological progress to feed 9 billion people (the estimated world population by 2050) in such circumstances, and a huge amount of energy would probably be expended to cool down our immediate living environment.

Time matters a lot. Over centuries, populations might move and new civilizations might develop at the currently uninhabitable fringes of the planet. Canada and northern Russia might become densely populated. But a moment's reflection suggests that an attempt to make shifts on this scale within decades is not likely to succeed. As people struggle to cope with the warmer planet, many will want to move. Many might be desperate. Wars and mass migrations are therefore quite likely to be consequences of such major climate change.

Biodiversity in aggregate would take an enormous pounding. Separate from – but exacerbated by – global warming, the elimination of half the species on the planet by the end of this century is a real and very sobering possibility. Many more species are becoming marginal. Add in 4–6°C warming and the losses will be much greater. The speed and scale of the losses are already comparable to the great geological extinction episodes.[33] We might not easily get by with so few other species and the implicit collapse of ecosystems, short of large-scale genetic engineering. Some species will nonetheless thrive whilst others are being decimated, but it is a fair bet that the survivors would not necessarily all be benign from a human perspective. Medical science might not be able to tackle the consequences.

We do not know with any certainty whether 2°C will be the upper bound of global warming, but we do have reasonable grounds for thinking that if it is not, things do not look good for humans. The problem of climate change is very much about what has been called the 'fat tail': the small probability of something truly nasty. This is what can be referred to as catastrophe risk: events that would threaten the survival of significant parts of the human race. Whether climate change is an example is far from obvious. A large asteroid hit would probably be much worse. Indeed, in the geological past it caused mass extinction. Then there is generalized nuclear war, bioterrorism, and viruses. At the upper end of temperature projections, the detrimental impacts of climate change could be very large in ways that we cannot fully comprehend now. This gives rise to the question at the heart of this book: *what should we do now to mitigate the risk of a very bad outcome?*

Put this way, the difficulty with motivating people to act and the imperative of avoiding a catastrophe sit uncomfortably next to each other. How do you explain to people that, faced with their day-to-day struggle to better themselves and their families, they should forgo some of their consumption now in order to serve the possible interests of people in the future?

One approach is to over-egg the risks. Films about a climate Armageddon, and the tendency to attribute every extreme weather event to climate change, thereby stoking fears of mass flooding and deaths from heatwaves, are all part of a climate change alarmism which has so far failed to make much impact. Indeed, it may well be counterproductive: if the end is nigh, why not enjoy the last moments?

An alternative is to go down the Barroso route quoted above – pretend it is all good news and that mitigation is a great economic opportunity to 'surf' the new industrial revolution. The Stern Review plays to this second agenda by suggesting that the costs of mitigation may be low, and perhaps even positive. As we shall see, this approach is no more likely to be successful than the Armageddon approach, for the simple reason that it is unlikely to be true. People will have to pay – but which people? And which countries? This question of responsibility turns out to be anything but straightforward, and is yet another indication of why so little progress has been made. But before we can assign responsibility, we first need to identify the causes of the rising emissions.

CHAPTER 2

Why are emissions rising?

What causes the carbon emissions and, in particular, their growth? If man-made emissions are one of the proximate causes of global warming, what lies behind these emissions? Where do they come from? Why are they going ever upwards? Although to many the underlying causes may seem obvious, it turns out that understanding them radically changes the focus of policy that aims to limit – and then reverse – their growth. We have been looking in the wrong places, and it is therefore hardly surprising that climate policies have been poorly targeted and as a result have been ineffectual.

Coal, coal and coal

It is not hard to find the prime villain of the piece. It is the burning of fossil fuels – almost everyone knows this. What is less appreciated is that all fossil fuels are not equally bad and, of these, coal bears the lion's share of responsibility. Coal is worse than oil, and much worse than gas. It is a distinction that really matters.

The global demand for coal is awesome. Between 2000 and 2010, it grew by over 70%. China and India accounted for more than 90% of this growth. Coal's role goes right through the history of the transition from

275 ppm to almost 400 ppm. It is responsible for a lot of the stock of carbon now in the atmosphere. The Industrial Revolution was founded on coal, long before it became the primary fuel for electricity generation. It was the combination of coal as an energy source and the maritime capability supporting trade that made Britain the workshop of the world.

It is important to remember that for most of human history, labour was the primary energy source, augmented by draught animals, water power and wood (biomass). Agriculture was at the mercy of the elements, and food supply was both limited by, and dependent on, climate. Populations expanded in warmer periods and contracted in colder ones. The Roman Empire benefited from a benign climate, and the Vikings could thrive in the higher latitudes because they could grow their crops, even in Greenland. The return of much colder weather in Europe limited population and economic growth; to the extent that there was any growth at all, it was limited by the lack of labour power. When Karl Marx argued that labour is the basis of all value, and capitalism is a process of extracting the surplus from labour, he was really describing a pre-industrial society.

The Industrial Revolution undermined this dramatically. The constraint of labour supply was relaxed by new sources of energy. Fossil fuels opened up a new era in human history. It is hard to underestimate the revolutionary effect of moving from labour to fossil fuels. Suddenly 'horse power' took on an altogether new meaning: coal could generate an exponentially larger number of horse-equivalents, and the new and improved technologies that it facilitated, like large-scale iron and steel production, and later railways and significant land transport, meant that agricultural productivity could rise. Without fossil fuels, markets would always be predominantly local; trade would remain restricted to sail; and self-sufficiency and local 'independence' in food supplies would remain at a premium. Cities would be small and rare.

The growth of coal changed all this. In Britain, coal output rose from 16 million tonnes in 1800 to 204 million tonnes in 1950. So great was the demand that by the late nineteenth century there were fears that it would run out. In industrial Britain, a leading economist of the day, William

Stanley Jevons, wrote about what we would now regard as 'peak coal' in his 1866 book *The Coal Question*.[1] He argued that national security depended upon security of coal supply, and that it was urgent for Britain to find new sources of provision. Coal for Jevons was 'the material energy of the country'.[2] The 'supposed substitutes for coal' are each considered and dismissed: 'it is useless to think of substituting any other kind of fuel for coal'.[3] The reality was for Jevons that coal supplies were limited and Britain should prepare itself for the consequence: 'Britain may contract to her former littleness'.[4] When Jevons wrote about an imminent scarcity of supply, coal was about to embark on a new and massive expansion as the primary fuel for generating electricity, a role it continues to fulfil today.

Coal, from an environmental perspective, is really dirty stuff. From its early development in the Industrial Revolution it has been associated with air pollution, slag heaps, water pollution and serious health impacts. Anyone who works in a coal mine is likely to have his or her health damaged in some way. For many, it is one of the more certain ways of reducing life expectancy. Lung disease is the most obvious consequence. To the occasional focus on dramatic mining tragedies are added the more frequent, lower-profile individual deaths and injuries. Compared with any other fossil fuel (and indeed nuclear), coal is the deadliest for its workers.[5]

The pollution starts as soon as the mine opens. Methane, a potent greenhouse gas, is released as the coal seams are opened up. Lots of energy is required to get the coal out of the mines. This was originally driven largely by human labour, with literally millions employed, but electricity gradually made a major impact, and mining moved on from being one of the biggest employers to being one of the largest industrial electricity consumers. Open-cast mining adds a new layer of environmental problems. It literally scores the landscape, leaving destruction in its path. Of India's coal production, 90% is open-cast, and it comes with a host of pollution problems.[6]

Much of the coal then needs to be transported to industry and power stations. Because coal is so bulky, and hence needs a lot of energy to move it around, power stations tend to be located near coal mines, and therefore not necessarily near the final electricity consumers. There are consequent losses of electricity in transmission. (Gas, on the other hand, is cheap to transport, and can therefore be turned into electricity nearer the market.) Once the coal gets to the power station (or industrial factory) it is often stored in big heaps. These 'rot', giving off gases and losing thermal efficiency. They also tend to have a lot of radioactivity.

The coal is then burnt in power stations. Here the pollution really gets going: much of the energy is lost in wasted heat, and large quantities of water are typically needed for cooling, resulting in warmer water in rivers and lakes. Combustion produces a cocktail of polluting gases – notably sulphur oxides (SOx), nitrogen oxides (NOx), carbon monoxide (CO) and carbon dioxide (CO_2). The SOx in the atmosphere falls as acid rain, with very damaging consequences for forests and lakes.

This led to the first large-scale sulphur emissions trading system in the US (an important and influential forerunner of the EU Emissions Trading Scheme (EU ETS) and the Large Combustion Plant Directive (LCPD) regime in Europe.[7] Whether through trading or regulation, the effect was to require some form of filtering and capturing of the SOx, typically by flue-gas desulphurization (FGD) technologies. This has had the incidental effect of further reducing thermal efficiency, and therefore increasing the amount of carbon pollution per unit of electricity produced, as more coal is required per unit. At considerable cost, acid rain in the US and Europe has been limited, although it remains a problem, and indeed these non-carbon forms of coal pollution remain of considerable concern not only in the US and Europe but also throughout the developing world.

At the domestic end, coal has played a big part in household heating and hot water supplies. For the first half of the twentieth century, coal, along with wood, was a primary fuel for heating the home in northern Europe. Open coal fires make even coal power stations look efficient by comparison. Since houses tend to be grouped together in towns and cities, the impact on human health of the resulting pollution was concentrated too. Coal fires caused smog when high-pressure weather systems were dominant – and smog killed in the thousands. The impacts were so great that open coal fires were banned in many urban areas, giving rise to clean-air legislation in most developed countries.

Oil and gas and their pollution

Other fossil fuels cause pollution too, but not on the scale of coal. Oil comes nearest, and oil pollution starts with the development of oil fields, including the impacts of spills and the flaring-off of associated gases, continuing with the process of refining and shipping, and final use in industrial plants, heating systems, electricity generation and transport.

The history of the oil industry is punctuated with disasters. The early years of the industry, notably around Baku in what is now Azerbaijan, is the stuff of pollution nightmares. Pollution on land and at sea was common. Refineries have let loose a range of emissions, and are themselves energy-intensive. There have been large-scale disasters at sea, with the *Torrey Canyon* shipwreck off the coast of south-west England in 1967, the *Exxon Valdez* running aground off the coast of Alaska in 1989, and the *Deepwater Horizon* blow-out in the Gulf of Mexico in 2010 standing out as stark examples. But to these can be added a host of smaller spills and leaks, and the flushing of tanks at sea. We are all familiar with the harrowing coverage of dying seabirds bogged down in oil slicks, but the true legacy is more enduring and probably impossible to calculate. Who can really say how much damage has been done to the Gulf of Mexico's ecosystem? Or to the pristine waters in Prince William Sound in Alaska?

The use of petrol and diesel in cars and trucks adds yet another dimension, from carbon emissions to other pollutants and particles – enough to require stringent regulation. Lead and sulphur in petrol have been substantially reduced, but the bulk of the pollution remains, to be inhaled primarily by city dwellers. Cars in cities contribute to smog, and the particulates emitted can cause breathing difficulties and lung disease. Across the world air quality in cities is a major health problem. China in particular has suffered the consequences, as witnessed in the run-up to the 2008 Beijing Olympics when air pollution was a major worry for the athletes. A further factor in the relative pollution roles of oil and coal is that oil is much more efficient in use. Most of the energy potential of coal is lost in the production process from mining through to the thermally inefficient power station.

Oil remains a dominating fuel in transport, and the projected growth in cars in developing countries is awesome. One study from 2008 suggests that the number of cars worldwide will increase by 2.3 billion between 2005 and 2050 and, of these, 1.9 billion will be in emerging and developing countries.[8] This is a simple extrapolation from an assumption that car ownership kicks in when per-capita annual income reaches $5000, and takes off above this level.

Aviation fuel provides a serious source of transport emissions, and it is all the more damaging by being released into the atmosphere at great height. The effects are of the order of two-to-four times those of land-based emissions.[9] The growth of aviation over the coming decades is likely to be strong, again especially in developing countries. In its 2011–16 12th Five-Year Plan, China has plans for 45 new airports. Significant world aviation growth is predicted elsewhere too. Over the next two decades growth of aviation passenger miles is expected to range between 4–6% per annum, with much higher growth in China and India. It is likely to remain oil-intensive.

Gas is more benign than both coal and oil, but it is not innocent. It does not cause acid rain or most of the other non-carbon pollution of coal, and nor do gas leaks produce slicks like oil. Yet it is potentially highly polluting since methane is a potent greenhouse gas – about four times more so than CO_2. It does, however, have the advantage of being short-lived in the atmosphere compared with CO_2, which will be present well into the next century and beyond, whereas today's methane emissions will be gone by 2050. Flared gas from oil wells and leaking pipelines are problematic. Shale gas adds a further layer of environmental problems, since there is associated pollution not just from the methane leakage, but also from the chemicals used to extract it and the implications for water supplies, to which we return in Chapter 10.

All these fossil fuels are bad news from a pollution perspective, but they are not *equally* bad. As the chart below indicates, a crucial distinction is in their associated carbon emissions, and here gas for electricity generation turns out to be about half as polluting as coal, and electricity is increasingly the way we choose to use energy. (Oil is used for electricity generation in developing countries, and especially in the Middle East.) This turns out to be a crucial factor in assessing the policy options.

The ever-upward trajectory of coal-burn

Given the pollution impacts, the burning of coal should be the most important immediate focus of attention in climate change mitigation. Although we need to be careful about comparing a change in a stock (of CO_2) and a flow (of coal burned), the graph opposite nevertheless illustrates a general point: the increase in coal-burn has been associated

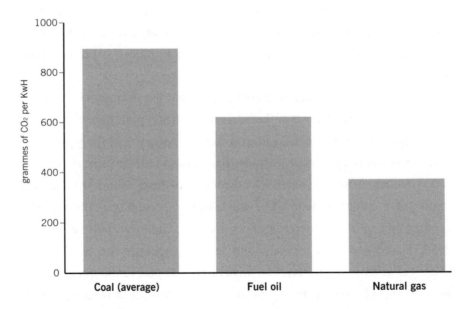

CO_2 emissions: grammes of CO_2/kilowatt hour of electricity generated

Source: International Energy Agency, 'CO_2 Emissions from Fuel Combustion: 2011 Highlights'

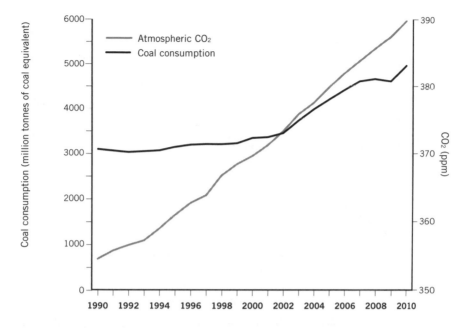

Coal consumption (million tonnes) versus atmospheric concentration of CO_2 (ppm)

Source: www.esrl.noaa.gov/gmd/ccgg/trends, BP statistical review of world energy 2011. Coal refers to commercial solid fuels only – i.e., bituminous coal and anthracite (hard coal), and lignite and brown (sub-bituminous) coal

with the increase in the stock of carbon in the atmosphere, especially since 2000. The flow has added to the stock.

The emissions path is ever upwards, and the coal path very roughly maps it. The only blip in the coal numbers comes after the economic crisis that began in 2007 and the global recession that followed, but the bounce-back was quick, and the trend pattern has returned. Furthermore, the blip was largely a phenomenon of the developed world. In developing countries, notably China and India, there was no serious setback to the rise of coal.

Projecting this coal-burn forward in terms of primary demand, the IEA 'World Economic Outlook 2011' provides three scenarios: current policies; new policies; and 450. 'New policies' represent the full implementation of proposed actions, and '450' is what would have to happen to limit climate

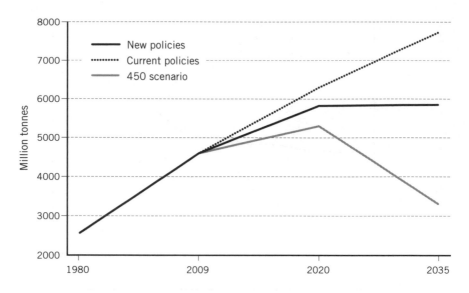

World primary coal demand by scenario (million tonnes)

Source: IEA, 'World Economic Outlook 2011'. Primary coal includes: hard coal (coking and steam coal), brown coal and peat

change to around 2°C. The graph above demonstrates just how big a challenge the 2° represents – and why in practice it is never likely to happen.

The China factor

Behind this global trend in the coal-burn lies the industrialization of one particular country – China. It is not the only coal-driven economy, but its sheer size and growth rates make it stand out. Its growth path over the last two decades has been phenomenal – with an annualized growth rate of over 7%. At this rate its economy doubles roughly every ten years, so it has doubled twice since the start of the Kyoto baseline in 1990, and on current trends it will double again by 2020 – the prospective date by which carbon targets may be put in place if all goes to plan following the Durban conference. If it continues on this growth path it will be 16 times bigger by 2050 – the target date for global decarbonization.

This extraordinary growth path is due to a combination of abundant cheap labour and – until very recently – abundant domestic coal. The flow of labour from the countryside has given China a competitive advantage, as has its weaker pollution controls and the absence of a carbon price and other climate change mitigation policies. China's electricity is around 80% coal-generated, and its share of the world's coal-burn is now close to 50%. Its energy-intensive industries – such as iron and steel, chemicals, ship-building, cement and fertilizers – rely on coal too.

The numbers are simply staggering. As the IEA's 2011 medium-term market report for coal puts it:

China's share in global coal production is almost four times that of Saudi Arabia's production of oil. Its share in global coal consumption is more than twice that of the demand for oil in the United States. Overall, the Chinese domestic coal market is more than three times the entire international coal trade.[10]

Coal-burn has historically been supported by its indigenous coal production. China's mines met demand up until 2009, and then it started importing. These imports come primarily from Australia and Indonesia (see table below).

	2000	2005	2010
Total imports	2.1	26.2	176.3
Australia	1.5	6.7	37.0
Canada	0	1.2	5.6
US	0	0	4.4
Indonesia	0.1	2.3	67.3
Russia	0	0.9	10.2
Vietnam	0	14.7	18.1
Other	0.5	0.4	3.9

Sources of China's coal imports (million tonnes)

Source: International Energy Agency, 'Coal Information, 2011'

Its coal mining industry has traditionally been diverse, with lots of small mines. In recent years, there has been significant consolidation, and production is increasingly moving away from the east coast, with the result that coal is now frequently transported up to 500km by rail to the industrialized centres of demand.

China's electricity industry struggles to keep up with demand growth, and not surprisingly there is a large-scale investment programme. Coal-burn in China's power stations increased by 13% in 2010 alone. Of the 2300 coal-fired power stations in the world, 620 are in China. The expansion plans for the rest of this decade will push the total towards 1000 GW (1 GW, or gigawatt, is roughly equal to a couple of medium sized coal power stations). This represents around 70 GW *extra* per annum, or around two large power stations *per week*. Put another way, each year China adds capacity of around the size of the entire installed capacity of the British electricity system; and since the bulk in China is coal, and only about 30% of British generation is coal, this represents adding more than twice the pollution of Britain's electricity industry *per annum*. The resulting emissions therefore represent a significant proportion of the extra annual emissions of carbon at the global level. This is the reality that climate change policy needs to address.

Projecting forward this growth of China's coal-burn is dependent on a list of variables. Yet it is not that hard. There are two ways of looking at China's coal future. The first is to take the IEA's medium-term forecasts (2011–16), and the second is to take China's own 12th Five-Year Plan.

The IEA 2011 coal report notes that China currently has 90 GW of coal-fired capacity under construction, with an additional 230 GW planned by 2016 (see chart opposite).[11] This is assuming a lower growth rate for the coal-burn in power stations of 5% per annum, compared with 9% per annum between 2005 and 2010. As a crude extrapolation, we could assume that in the subsequent five-year period, the same pattern is followed, and hence we could double the numbers for 2020, when a new political global agreement is due to take effect. Thus we might be contemplating up to 460 GW of new coal in this decade. This is more than the

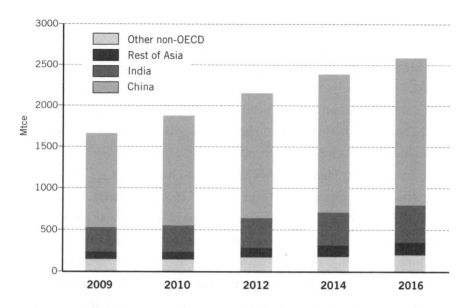

Projection of coal demand in power generation for non-OECD countries (million tonnes of coal equivalent)

Source: OECD and International Energy Agency, 'Medium-term Coal Market Report 2011: Market Trends and Projections to 2016', © OECD/International Energy Agency, 2011, Figure 9, p. 32

entire current coal capacity installed in Europe and completely dwarfs Europe's renewables ambitions over the same period.

However, this simplistic extrapolation would fail to take account of attempts by China to increase its energy efficiency and diversify away from coal. The alternative approach is to take the Chinese 12th Five-Year Plan seriously and assume it is achieved. This projects a major effort to reduce its energy intensity by 16% between 2011 and 2015, and to promote renewables and nuclear power in an attempt to diversify away from coal.[12] But it all needs a sense of proportion – the plan is to expand renewables to 2.6% by 2015. More significant is the push on natural gas and eventually shale gas too, with a doubling of its share of energy envisaged by 2015.

Even if all the Plan is achieved, the share of coal in primary energy consumption will fall from 70% to 63%. This is a falling share of a sharply rising number – and as a result it makes little significant difference to the trend in aggregate emissions. The effects might reduce the 460 GW to

400 GW, or even 350 GW at the limit. Even in this more benign scenario, it is still an awful lot of new coal power stations, which could operate for another 40 years or more. They will still potentially be on the Chinese electricity system in 2050. What is therefore needed – and quickly – is something altogether more radical: to head off this rising coal-burn.

The motives for China's new energy policy in the 12th Five-Year Plan are numerous. They can be presented to the world as a serious intention to address climate change, although closer to home they are driven by a combination of import dependency, price, competitiveness with lower-labour-cost neighbours, and the development of new (infant) industries to take advantage of developed countries' subsidies for renewables.

China's 'economic miracle' has, as noted above, been based upon the supply of cheap labour from the countryside plus abundant domestic coal. It now inevitably finds itself exposed to *cheaper* labour from its neighbours, and it has resorted to importing coal to meet demand. To compete in export markets, it needs to reduce other costs to offset labour-cost increases. Energy efficiency is an obvious opportunity. To address security of supply, it needs to diversify from its 80% coal dependency for electricity generation. This means other fossil fuels – notably gas – and nuclear, with 24 reactors under construction. Finally, with depressed consumer demand in the US and Europe (together around 50% of global GDP), China needs new export markets. The subsidies to wind and solar in Europe and the US are obvious targets.

The 12th Plan of course may not be delivered. On the one hand, the economic growth may not materialize (see below). On the other hand, plans and outcomes depend on a combination of power, direction, force with directed investment, and delivery. Economists have long doubted the efficacy of state 'plans', and been sceptical of state direction. Whilst authoritarian regimes can more easily trample on local opposition and land rights, the ability to translate targets into new assets is open to serious question. Doubts also exist about whether the reported numbers on progress are real or, like the experience of the Soviet regime, whether they are produced to fit the objectives.

This matters because it is more likely that the coal stations will be built than nuclear and renewables. Coal is a well-understood and mature technology for electricity. Nuclear has inherent technical challenges, whilst renewables are expensive relative to coal in the current context of coal prices. China may be able to withstand the concerns raised by the 2011 Fukushima accident in nearby Japan, but even it announced a temporary halt to nuclear construction. Internal opposition to nuclear development remains.

In Part Three we consider how this Chinese coal juggernaut may be turned around. But before we do, we need to recognize that although China dominates when it comes to coal, it is not alone. Whilst China's economic miracle has gained the media's attention, other developing countries are doing their bit to increase coal-based emissions. India comes second, with high economic growth, supported by coal.

India's demand for coal has been rising sharply in recent years. Again it is driven by electricity – with just under 70% being coal-based. India has lots of new coal-fired power stations under construction and plans a 30% increase in the coal burned in power stations by 2016. Given that around a quarter of its population has no access to electricity, the combination of extending the supply of electricity and its population growth makes the demand for coal for electricity very robust. It is likely to add at least one new power station per week over the period. So India and China together add up to three large coal-fired power stations *per week*. To put this in perspective in terms of renewables in Europe, assume that this amounts to 1 GW. It would take 200 very large 5 megawatt (MW) wind turbines to match this *per week* if they had the same load factor. But they don't. They are intermittent, and 2 MW is a better number for the average size. Crudely, assume they have half the load factor. This is now 1000 wind turbines per week – or 52,000 per year – just to keep pace with India and China's extra coal-burn. Whether these numbers are out by a few thousand a year either way (and they are very generous to wind) does not much matter: it is the sheer scale of what is going on that matters.

Unlike China, India does not have such ambitious plans for nuclear, renewables or energy efficiency. This is *not* to suggest that these will not play a part – they will. India plans more nuclear power stations (it has six under construction, compared with 24 in China),[13] and is expanding wind and solar. But it starts from a lower base of energy use per capita, and hence it is further down the curve of the energy growth path. Thus the sorts of policy changes envisaged for China may be a decade away for India.

After China and India, there are a host of other countries with significant coal production and coal-burn for electricity generation, as the selection in the table below illustrates.

South Africa	93
Poland	90
China	79
Australia	76
Kazakhstan	70
India	69
Israel	63
Czech Republic	56
Morocco	55
Greece	55
US	45
Germany	44

Coal in electricity generation for selected countries, 2010 (%)

Source: World Coal Association Coal Statistics, 2010, available at www.worldcoal.org/resources/coal-statistics/

South Africa provides a good example of the challenges from a climate change perspective. It is a coal-based, mining economy, and is building more coal-fired power stations. 93% of its electricity is generated from coal. It ranks in the top 20 countries for emissions. Just one coal-to-liquids (CTL) plant at Secunda, a legacy of the apartheid years when it faced an oil embargo, ranks amongst the top single sources of carbon emissions in the world – with more than the total emissions of half the world's countries. They do not propose to close it any time soon. Hosting the Durban climate conference at the end of 2011 provided an interesting illustration of the difference between the ambitions of the parties and reality. South African emissions will grow strongly. It will burn more coal, and it plans several more coal-fired power stations.

By contrast, the developed world has arguably peaked in its coal demand. There are lots of reasons for this, including: low economic growth, low energy demand growth, regulation, and carbon prices. Consequently there has been little need (outside Germany) to build new coal-fired power stations, and where more electricity generation capacity is needed, gas has become very competitive. But the lack of a rising trend should not disguise the scale of coal-based electricity capacity in the developed world. The US figures strongly, with around 340 GW of coal-based capacity, and has been producing around 45% of its electricity from coal. However, unlike China and India, the competitive challenge of gas is already beginning to make serious inroads into the coal share of US electricity generation, and a programme of closures is under way. Many are old, having been built in the 1960s and 1970s.

We return to these crucial gas-substitution opportunities in Chapter 11, but we should note here that falling demand for energy in developed countries does not mean the moderating demand for carbon- (and coal-) intensively produced products. As we shall see in the next chapter, the deindustrialization that accounts for falling demand in developed countries has been replaced by imports of carbon-intensive products from countries like China. Because China burns the coal, it does not follow that it is all for China's own consumption.

There is plenty of coal left

The sheer scale of the coal-burn might lead to the expectation that supplies will eventually run out as Jevons predicted in the nineteenth century. Sadly, here too there is no silver lining. There is no 'peak coal' supply in sight. On the contrary, there are probably several centuries of reserves left. When photosynthesis captured carbon from the atmosphere millions of years ago and created the coal reserves we are now exploiting, it did so on a massive scale – just as we are transferring that carbon back into the atmosphere on a massive scale now. In the process it put a lot of oxygen into the atmosphere – indeed so much that megafauna could flourish.[14]

Coal reserve data produced by international agencies and national governments tends to be conservative: there is so much of the stuff that it is hardly worth the cost of accurate surveys. The US records the largest coal reserves in the world, and has enough recoverable domestic coal reserves at current mining levels to last 222 years.[15] Russia and China have the next largest known reserves. India has lots too.

Not only are coal reserves super-abundant; they are also widely distributed. Until recently, most major coal-burning countries have been largely self-sufficient, or been able to rely on near (and friendly) neighbours. China, India, the US, Poland and Russia all fall into this category, as once did Britain, Germany and even France.

Now the major players – China and India in particular – rely on imports *at the margin*. The corollary of these imports is the exports of countries like Australia, Indonesia, and Russia, to be augmented in the future by new players like Mongolia and Madagascar.[16]

In summary, coal is environmentally awful. It is the main driver of emissions growth; it is going to go on playing this role, especially in China and India; and there is plenty of it to keep the pace of increase in coal-burn going for decades, and possibly centuries. The force driving this increasing coal-burn is economic growth in the key developing countries, which is our second (related) cause of growing emissions.

Economic growth and ever more consumption

Why is coal-burn going ever upwards? The answer is economic growth. We cannot know what the rate of economic growth will turn out to be over the coming decades. There is no smooth and continuous growth process and, despite the burgeoning business of economic forecasting, even a few months out and the scale of error can be (and often is) large. The best we can do is sketch out possible scenarios.

The history of economic growth is one of punctuated bursts. The twentieth century overall witnessed growth rates of around 2% per annum in the developed world; and looking further into the past, the British and US economies have managed between 1 and 2% per annum over the whole period since the Industrial Revolution.

For developed countries, 1945–70 was the great growth period. 3% growth rates in Britain and the US, and much higher rates in Germany and Japan, led to a massive increase in energy demand. Germany and especially Japan emerged from the devastation of the Second World War as economic powerhouses, with Japan eventually becoming the second-largest economy in the world (after the US). For Britain, 3% per annum economic growth fed through to 7% growth in electricity demand, and across the developed world coal was overwhelmingly the primary fuel for electricity, and oil for transport.

The 1970s turned out to be an energy-related interruption in the growth path, creating crises in Europe and the US, as oil prices first quadrupled and then, at the end of the decade, doubled with the Iranian revolution. Stagflation pervaded; the Bretton Woods exchange rate system collapsed; and the International Monetary Fund (IMF) was needed to rescue the British economy.

The end of the 1970s saw expectations of ever-increasing oil prices. This new conventional wisdom encouraged all sorts of energy-diversification strategies – then, as now. In fact, the oil price collapsed in the mid-1980s, and has since remained for the most part below the 1979 peak price in real terms. Growth resumed; and Margaret Thatcher and Ronald Reagan's market-based politics unfolded into economic policy. In the 1980s the economic future looked very Japanese – in the way that today many commentators see the future as Chinese. In 1989 Japan's economy – based on exports to the US, financed by lending the US the money to cover its trade deficit, and in turn financed by Japan's high savings rate – collapsed.

It was to remain in stagnation for over two decades. Its stock market had peaked at 39,000 in 1989: in early 2012 it was still below 9,000.

In growth terms, Japan was eventually replaced by China, on a very similar model: based upon and supported by exports and high savings, and underpinned by purchases of US Treasury bonds by way of financing the US purchases of Chinese goods. China's growth since the 1980s has been spectacular, remarkably without much interruption. From a climate change perspective a key question is whether it will continue.

The starting point is to project current growth rates of around 7–10% per annum. The implication is dramatic for the period through to 2050. As already noted – and cannot be repeated too often – at 7% per annum, China's economy would double by 2020, and double again every decade. The result is both staggering and, from a climate change perspective, potentially disastrous. It would have caught up with US current GDP per head consumption, and for a lot more people – 2 billion in total. India's position would not be much different. This means that by 2050 there would be around 4 billion people all at current US-style consumption rates (as well as roughly a billion in the US and Europe).

The 12th Five-Year Plan envisages a slight easing of the very high growth rates of the last decade, and recent official revisions suggest that in 2013 it will slow from more than 8% to around 7% per annum. No major shocks are, of course, officially foreseen. Its export model, as noted, has been built on cheap labour and cheap energy, and the former looks less sustainable going forward as the authoritarian regime continues to offer material progress through higher standards of living in exchange for acquiescence with and obedience to the State. Its export markets have contracted as the great late-twentieth-century boom in the developed countries has come to an end.

Can China switch gradually from an export to a domestic consumer-led growth model? Can it make the transition from rapid catch-up growth to developed country rates – from 7–10% to 2–3% without political upheaval and reform? Many aspects of such a transition would prove challenging. The rise in consumption means a fall in savings and hence a fall in the financial surpluses. The investment rate of 50% GDP will have to fall accordingly. Current infrastructure investment may prove to have been wasteful, and the impact of external shocks may turn out to be difficult to absorb. On the other hand, energy prices may be more benign, given plentiful coal supplies and the prospect of shale gas.

Ultimately it is not a question that can be answered by forecasting and prediction. Many argue that open societies, with freedom of expression and democracy, are more innovative, and that planning is open to the sorts of mistakes made in the former Soviet Union. From a climate perspective, a collapse in China similar to that of Japan in 1989 would provide considerable relief for the climate. It would probably be the single most important mitigation factor. But it should not be banked on. Hence whatever the short-term shocks and the vagaries of the business cycles, we have to craft climate policy on the *assumption* that the growth of China continues, with all its terrible environmental implications.

In the US and Europe, growth is likely to be more moderate, and less energy-intensive. Plodding along at 2 3% per annum may look tame by China's standards, but the power of cumulative growth rates translates this growth into a doubling every 20–25 years, and hence a potential quadrupling by 2050. And this is the rate of growth that delivered the current US and European economies in the twentieth century. Whilst current growth rates are grim, as with China we should not confuse short-term shocks and fluctuations with a long-run trend. It is the latter that matters for climate change.

To get a grip on what such economic growth around the world might mean, imagine everyone in the developed world having twice (or even four times) as much to spend. What would they spend it on? More travel and holidays? Air conditioning? A host of energy-using devices?

It is hard to imagine that such a rise in consumption would not translate into a significant rise in the demand for energy. The composition might change, and energy efficiency will make inroads, but the underlying story remains the same. It is pure fantasy to imagine that in such a growth scenario global energy demand might fall. Yet, as we shall see, it is a fantasy that is at the heart of many climate change policies, notably in Europe. Chapter 5 explores some of the energy efficiency myths built around the assumption that this sort of economic growth might be combined with little or no growth in energy demand.

Population growth

Completing the picture of the causes of climate change, behind the growth of emissions lies the coal, and behind the coal lies the economic growth, and behind the economic growth lies the increase in population. Between now and 2050, the world's population is expected to grow from its current

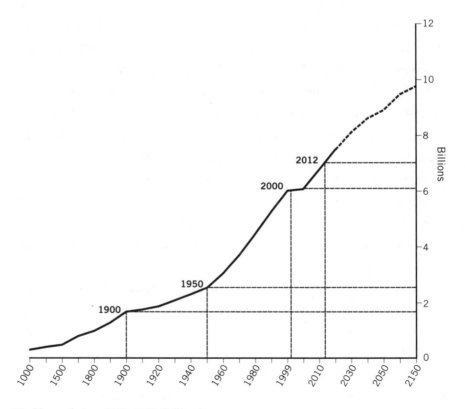

World population, 1000–2150 (billions)

Source: United Nations, 'The World at Six Billion', Department of Economic and Social Affairs, Population Division, October 1999, and United Nations, '2010 Revision of the World Population Prospects', 2011

level of 7 billion to 9 or 10 billion – or in terms of the Kyoto baseline in 1990, by 3 billion from 1990 to 2050. At the upper end this increase is more than the entire world population in 1950 (2.5 billion) as the graph opposite indicates.

The history of population growth is truly extraordinary. For most of human history, the global population remained fairly static. Numbers were limited by the dependency on human labour to provide energy, augmented by animals. It was around 1 billion in 1800 – just over double the world's population in 1300, and hence it took 500 years to double. Malthusian logic kept a lid on population size, in the absence of medicine and in the presence of the influences of weather and climate. A swing in climate – for example, after 1300 when the north European Little Ice Age began – exposed populations to malnutrition, leading to epidemics like the plague, which culled numbers further. In the fourteenth century, the Black Death may have reduced the European population by a third.

What changed that static world of physical constraints was the combination of fossil fuel energy, machines, chemical fertilizers and modern medicine. The Industrial Revolution relieved the constraints of manpower and horsepower with mechanization, tractors and nitrogen. Coal-generated steam, followed by electricity and oil, opened up markets. Modern medicine and hygiene cut the death rate, so more people could be fed and they lived longer.

The impact of these technological advances began in the nineteenth century, but took off in the twentieth century. In 1900 the world population was around 1.7 billion. By 2000 it had reached 6 billion, with the last billion added in the last 12 years of the twentieth century. By 2050, it will, as noted above, probably be around 9 billion. It is hard to grasp the reality of these numbers. The majority of people who have ever lived are alive today.

This population growth has been accompanied by an explosion in consumption, as living standards have risen with economic growth. These new extra people will add their extra demand – and their consumption.

They will be concentrated in China, India and Africa. The Chinese and Indian populations are now over 1 billion each. By 2050 they will probably be 2 billion each. An extra billion will be added in Africa, bringing it to 2 billion as well.

All these people will require food, heating, and housing. These requirements cannot be delivered without energy, and hence they represent a wall of new energy demand even if they are all poor. But if they aspire to the standards of living currently enjoyed by developed countries, the *extra* energy demand would equal half of *current* global energy demand – on top of the growth of energy demand to meet the higher consumption of the other 6 billion.

Population optimists point to the claim that as people get richer, their birth rate declines. They shift from large to small families. At the limit – for example, in Germany – they have too few children to replace those who die, and the population starts to level out and then decline. They go through a demographic transition.

China is increasingly beginning to look like this: the one-child policy means that the population will age over coming decades. This creates its own problems, but again we need to separate out hope from reality. Even if the world's population levels out by 2050, what happens between now and then is still a new wave of consumption, and with it, energy demand. It is more than enough to create the extra demand which, if met by fossil fuel supplies (and especially coal), will most likely trigger climate change above the 2°C benchmark. Short of a major catastrophe – a lethal infectious disease, an asteroid collision or a major nuclear war – these extra people are going to materialize. Lethal infectious diseases on a very large scale are in evolutionary terms inefficient: the trick is to keep the host alive. Asteroid collisions really are rare, although perhaps we should take them seriously.[17] Nuclear war is an ever-present threat, although perhaps surprisingly it has not happened since Hiroshima, and if it did it is far from clear that it would kill billions. So we need to assume that the total will be 9 billion, and we have to plan for this.

Any serious attempt to address climate change first has to address these facts. They are not good: despite all the international efforts, the coal-burn keeps going up, and with it emissions. This trend is deep-rooted: it is underpinned by economic growth, and in turn by population growth. The climate change challenge is therefore how to get off the coal hook, whilst at the same time accommodating much more consumption and many more people. Oil will have to be phased out too. It is a big ask.

CHAPTER 3

Who is to blame?

Who is to blame for the rising emissions? Who is responsible? Is it China, and to a lesser extent India? China burns the coal; the coal leads to emissions; the emissions cause global warming; and therefore it is easy to conclude that China is a major cause of global warming.

But whilst China may provide a convenient scapegoat, assigning responsibility is an altogether more complex matter. This requires an ethical dimension. Responsibility is not just about causality, but also about rights and entitlements. Hence we need a framework for deciding not only who *caused* the carbon emissions, but also who is *entitled* to make these emissions – in the past, now, and in the future – and who should bear the burden of reducing them. The ethics of climate change require judgements to be made about the relative claims of different individuals, groups and countries.

There are at least three ways of thinking about this. We can ask: who put the carbon into the atmosphere; what would be fair; or who is responsible for creating all the carbon consumption now? As we shall see, the answers all turn out to be rather similar.

Who put the carbon into the atmosphere?

An obvious approach to assigning responsibility is to consider who did the polluting, and to apply the polluter-pays principle. As a concept, however, there are some considerable difficulties – the polluter is not always easily identified (and is often dead); the damage can be uncertain (and often undetected when the pollution takes place); and even when the polluter can be identified and we know the damage, it is not even clear that it is always the polluter who can (or even should) pay.

All of these difficulties arise in the climate change context. The atmosphere is a public rather than a private good. No one owns it in terms of identifiable individual property rights. Not even emissions permits create ownership of the atmosphere. Its pollution is therefore a common activity – and hence it is a classic example of the tragedy of the commons, and will be over-exploited as a result.[1]

If responsibility arises because of *ex ante intentions* rather than *ex post outcomes*, then it is hard to claim that over the last couple of centuries, anyone – or any corporate entity or government – intended or knowingly *caused* climate change. People living today clearly did not cause the past emissions.

Nevertheless, and notwithstanding these difficulties, it is reasonable for developing countries to argue that whilst they might be causing much of the increase in emissions *now*, the stock of carbon put into the atmosphere since the Industrial Revolution has been largely caused by the US, Russia and Europe. It is also their current populations who enjoy the capital stock that has developed in part as a result of burning coal and other fossil fuels. The current living standards in developed countries are the outcome of all that pollution that is reflected in the rising stock of carbon in the atmosphere.

Precisely how much has been emitted by each country and each company is very hard to estimate. There were very few records of emissions of the relevant gases from particular industrial plants until very recently, and even these are subject to uncertainty, since there are obvious incentives to misrepresent the data and present a biased interpretation. What level of emissions came from Stalin's forced industrialization? From the two world wars in the twentieth century? From the Middle East wars?

Yet this uncertainty does not mean that nothing useful can be established. We know a great deal about industrial structures in the developed countries – about steel, chemicals and cement production, for example. We can also make educated guesses about the coal-, oil- and gas-burn. With these numbers to hand, an educated guess can also be made about historical responsibility for emissions.

For the purposes of an international agreement, rough answers are all that would be needed. They enable us to do two things: to identify the main historical polluting countries, and to take their GDPs and apply a carbon-based energy intensity to them. This puts the US, Russia, Britain, Germany, and a few other European countries in the frame. In this historical sense, they are the guilty parties.

With this responsibility established, it could be argued that any international agreement should bear down disproportionately on these countries. Although the Kyoto Protocol's separation of Annex I developed countries was supposed to roughly reflect this distribution, it did not do so in practice. The US did not ratify the Kyoto Protocol and has no emissions caps; Russia signed up only because it believed there would be little or no effect on its industries (which turned out to be a correct assumption); and the burden on Britain and Germany turned out to be weak, on account of the combination of deindustrialization (especially in the British case), the collapse of former East German industry, weak growth, and the switch from coal to gas. The recession has made the Kyoto targets easy.

Fairness

If one black mark against the developed countries is that they put most of the carbon in the atmosphere up there, another is that they also emit much more carbon per head. China may have a large population, but it is still relatively poor, and hence the aggregate emissions for China mask the fact that most citizens on average do not have the wealth to do the consuming of carbon-intensive goods that Americans and Europeans do.

The charts below present country comparisons of income per head and carbon emissions per head.

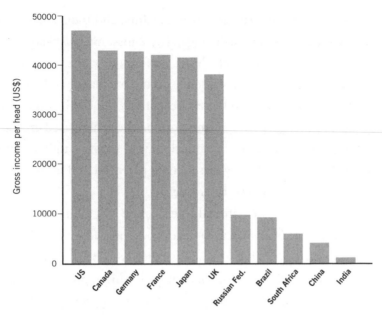

Income per head of population by country, 2010 (US$)

Source: World Bank GDP indicators http://data.worldbank.org

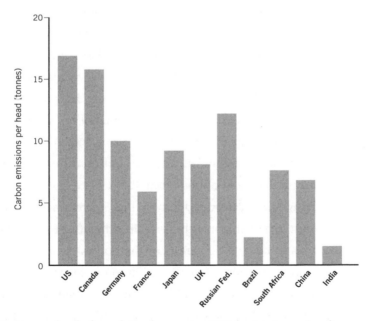

CO_2 emissions per head of population by country, 2010 (metric tonnes/year)

Source: European Commission Joint Research Centre background study on long term trend in global CO2 emissions, 2011

The income per head disparity between China and India and the major developed countries is obviously huge. The emissions per head numbers are a bit more mixed, and reflect China's rapid industrialization and South Africa's overwhelming coal dependency on the one hand, and the nuclear contribution in France and, until very recently, Japan on the other. (As we shall see later in this chapter, the emissions per head are based on national production of carbon, and consumption will paint a rather different picture, so that the match between income per head and emissions on a consumption basis is a much closer one.)

We can go beyond these national levels. Some rich Chinese individuals emit on a large scale; some poor Americans do not. Thus, a second, more fine-grained, way of looking at responsibility is to base it on each *individual's* pollution rather than that of countries. This approach is grounded in what might be described as 'deep impartiality' – a principle which finds expression in liberal political theories and centres on the idea that all people should be treated equally.

As a principle it has obvious intuitive appeal, but it is very radical. If each individual is equal, then *when* or *where* they live should be irrelevant. It follows that a radical redistribution of wealth is mandated, since it is patently obvious that some people live close to starvation now, whilst others are worth literally billions of dollars. These inequalities exist in countries, and between countries. The 'bottom billion' people – those living on less than a dollar a day – are not being treated impartially in respect of those in the 'top billion' incomes, or indeed the other five billion or so who make up the world's population.[2]

The problem with radical impartiality is that no one has convincingly worked out how a society that met this principle might actually work – and indeed how it could be in the interests of the poorest to have equality of outcomes. The political theorist John Rawls famously suggested that society should be organized such that inequalities could be justified only if they were to benefit the worst off.[3] In the social contract tradition of liberal political theorists from John Locke onwards, Rawls considered what principles of justice people would choose if they were abstracted out of society and placed behind 'a veil of ignorance' about the position in society they may personally occupy. Others have suggested that impartiality should refer not to outcomes but to some concept of 'equal chances' or 'equality of opportunity'.

The conceptual difficulties pile up as complexity is introduced and impartiality becomes part of a contested theory of justice. An equal opportunity to become unequal begs a host of questions about the starting line. Is it nature or nurture that explains different outcomes? What happens if my genes are less conducive to 'success' than yours? What if my parents are less attentive than yours?

None of these arguments suggests that we should not be concerned about the allocation of resources between individuals, or that *laissez-faire* will produce a good outcome. However, they do limit the application of impartiality considerably, and require – if they are to form part of the architecture of a credible climate change agreement – that they go with the grain of human nature. Ethics are not defined by human nature, but they are limited by it.

We do have altruistic tendencies and we do have a sense of justice as impartiality.[4] It is not only just but also practical to educate and persuade people to share some of their wealth with others less fortunate. And therefore the fact that emissions per head are much lower in developing countries – because they have less income per head – is relevant to apportioning carbon targets. But altruism starts at home, and diminishes as people are less connected with us – biologically, culturally and physically.[5] We are not behind a veil of ignorance: we are *in* society, and we occupy positions *within* society that are deeply relevant to our ethical judgements.

As Roger Scruton has eloquently argued, environmental concern begins with the family and the home, steps out into our local community, and then takes account of the nation state.[6] Loyalty and concern for international organizations are much weaker, and we tend to give through NGOs and special funds to help poverty relief, rather than to foreign governments. When it comes to governments, national interest dominates foreign policy.

The impartiality claim is too demanding in its pure form at a given point in time when considering the distribution of resources and wealth now. It is too strong a claim in respect of the radical idea of equal emissions rights per capita at the global level. It is also practically impossible to sort out and apply over the time period in which emissions need to be limited.

This approach to impartiality is, however, completely overshadowed by the claim made by Nicholas Stern in the Stern Review and elsewhere that we should be impartial *through time*.[7] He argues, taking his cue from Frank Ramsey and a school of (predominantly Cambridge) utilitarian thought, that people's welfare (expressed in utility terms) should be regarded as completely independent of the time period in which people live. Ramsey suggests that to discount utility – to place less value on future utility than current utility – is to show a 'lack of imagination'.[8] He argues that we should treat utility now and utility in a thousand years' time identically, and therefore we should care as much about people with whom we have only the vaguest tangency in the future – in a hundred, a thousand, or one hundred thousand years' time – as we do about our friends and family today. Stern agrees: we should, in the jargon, have a 'zero time-preference rate': 'if a future generation will be present, we suppose that it has the same claim on our ethical attention as the current one.'[9]

This is his very radical point of departure, and his conclusion about the economic case for urgent action conveniently depends upon it. Indeed, he goes on to make the remarkable claim in the Review that it is in effect a necessary condition for a concern about climate change – and thereby implicitly admits that his economic case for urgent action falls away without this zero time-preference rate. The approach is somewhat disingenuous: Stern states that 'many other economists and philosophers' agree with him, and then avoids any intermediate position by following this up with the statement: 'those who put little weight on the future (regardless of how living standards develop) would similarly show little concern for the problem of climate change'. Notice there is little room for those who show *quite a lot of concern* for the future, but not a zero time-preference rate. It results in a polarization of positions: you either agree with Stern's ethics, or you do not care about climate change.[10]

Stern does make one concession to what is an extreme demand on our imagination and our concerns: he does include the possibility that we will become extinct. Were this to happen, there obviously would be no more future generations to worry about. Stern rates the chance of human extinction surprisingly high, at 10% in the next 100 years, and he therefore adds 0.1% to the time-preference rate. So he has a small positive number to use in utility discounting. Given the relatively short period humans have existed, and given the damage we have already inflicted on the planet, it is not an unreasonable assumption that humans may become extinct. We may indeed even be 'suicidal'.[11] But, as discussed in Chapter 2, for climate change to be the cause of our extinction, some extreme assumptions about temperature rises and their impacts would have to be made.

If we combine this idea of impartiality for all current people – and hence equal rights to carbon emissions now – with Stern's zero time-preference rate for the future, we get the implausible idea that all people, everywhere and forever, should have an equal entitlement to emit. We should issue pollution rights or permits on an individual (equal) basis for ever, ignoring the national energy systems and energy policy decisions of countries. The close relationship between socialism and impartiality is apparent: radical impartiality is in essence an argument for radical equality. As such it has force – but the force of socialism in general rather than climate change in particular – and it is therefore vulnerable to the general critique of socialism as a concept.

With this in mind, it is no accident that many (but of course by no means all) environmentalists are on the political Left in their politics. The impartiality argument is a claim about 'fair shares' and hence redistribution, and redistribution is the territory of the Left rather than the Right. It stands in contrast to arguments based upon stewardship which conservatives advance. The mainstream Right starts with the current distribution and human nature as it is, and – from a moral focus on the home – considers how to husband the environment. In Scruton's terms, it is a conservative green philosophy that does not rely on an ideal distribution of income or wealth, or indeed the perfection of human nature. It is necessarily limited by its view of human nature.

The libertarian political Right, following Nozick,[12] also starts from the existing property rights, but argues that problems of pollution and externalities should be left to the parties to negotiate. The approach gained powerful intellectual support from Ronald Coase's bargaining theory.[13] Coase claimed that market failures, like pollution, were not really failures at all: the affected parties could bargain amongst themselves, on the basis of who had which property rights. The emphasis here is on ownership, and what to do in its absence, in examples like the oceans and the atmosphere. Rights – like pollution permits – should be created. It is a small step to emissions permits and trading, and the EU ETS, which we discuss in detail in Chapter 9. It is perhaps no accident that some in this camp prefer the fallback of pure climate change scepticism.

The wider political context is reflected in green politics. Green parties are almost everywhere of the Left, and often further to the Left than social democratic parties. Green parties tend to be suspicious of business, hostile to capitalism and markets, prefer planning and regulation, and disparage economists and economics. These approaches fit naturally with the idea of environmental equality. Recent 'solutions' promoted by Mayer Hillman (*How We Can Save the Planet*) and George Monbiot (*Heat*) provide examples of this planning approach.[14] Hillman describes 'fair shares' as 'the only way forward'. People would be given 'equal rations' of carbon; these rations should then be reduced. It would have to be 'mandatory'. Monbiot is similarly interventionist. He argues that only a 90% cut by 2030 will save us, and that it is the cowardice of politicians and the power of corporate lobbyists who stand in the way of radical measures which we should presumably be forced to adopt.

The policy of globally handing out permits per head flows from this approach.[15] This might lead to trade, with a permit price established at a global level as a result. Such trading could quickly lead to inequality – after all, that is what capitalism and markets do. Some people are more economically successful than others. So, if we are to be strictly impartial, trading would need to be limited or the outcomes redistributed. The result is not only *ex ante* carbon rationing, but also *ex post* redistribution. But even if it were only *ex ante* equality (as with the handing out of shares in state companies in eastern Europe after Communism collapsed), and the *ex post* inequality that emerged was tolerated, the sheer scale of the transfer of resources from the developed to the developing countries would be enormous. Over the time horizon in which emissions need to be reduced, this is not going to happen. Global socialism, whether desirable or not, is not going to solve global warming.

What is wrong with Stern's zero time preference mirrors what is wrong with impartiality as a general universal principle. Human nature is extremely partial: we focus close to home, and our concerns diminish as our horizons widen. Scruton is right about this. Education, moral philosophy and social engagement widen the domain of altruism. But we start with self-interest. Ramsay's claim can be turned on its head – it is not a 'lack of imagination', but rather a lack of an appreciation of human nature on Ramsey's – and Stern's – part.

The recognition of partiality has an implication for the attempts to address global warming. The impartiality approach suggests that we should address the problem for the good of *all*, now and in the future. Partiality suggests that we should address climate change primarily because it is in our *self*-interest to do so. If we do not, then all sorts of bad things will happen *to us* and *to our children*, and since the developed world has the resources, if it does not act, climate change will happen, and we will all suffer accordingly. Grandstanding on particular ethical principles has its place, but it is not going to transform people into followers of the advocates' preferred ethical principles in the near future – or ever. It is therefore not at all surprising that it has had no discernible influence on climate change policy. It is interesting academic scribbling, not a guide to practical policy.

Starting with self-interest does not mean that wider concerns about justice do not enter into our consideration. Self-interest extends to families, and children in particular, and this is the most immediate focus of attention and wider moral claims. The next generation is something that matters to us, and often much more than the one after that. Climate change is predicted to make a significant impact on today's children when they are adults: they should on average live until almost the end of this century, by which time warming may be between 3°C and 6°C.

If we can show that addressing climate change is a sensible strategy from a narrow basis of concern, then adding in wider concerns of justice provides a bonus. It also focuses attention on those measures that can be implemented *now* with low costs, taking as given the existing distribution of wealth. For example, a switch from coal- to gas-fired electricity generation has very low cost, reduces carbon emissions significantly along with lots of other immediate types of pollution, improves health and saves lives. None of the proposals advocated later on in Part Three relies on Ramsey's or Stern's radical impartiality.

The corollary is that if more costly mitigation strategies are chosen now, there is likely to be greater resistance, because they conflict more sharply with immediate self-interest. Indeed, this is precisely what has happened as some of the most expensive options have been chosen first, and customer resistance has grown as the effect on energy bills has been felt. The greater the costs, the more prominent the distributional issues. It is not an accident that those who advocate the most demanding ethical principles often have less concern about the costs and people's ability to adjust. Since they *know* the right answer – both in fact and morally – why not go straight from here to there? And if people do not share their moral zeal, well why not force them to take rations? It works in wartime, so why not in the 'carbon war'?

Giving human nature a central role, and gradually widening the domain of ethical concern, is an idea that derives from the Scottish enlightenment philosopher David Hume, rather than the utilitarianism that lies at the heart of Ramsey and Stern's ethics. Hume famously distinguished between 'ought' and 'is'. What 'ought' to be is not determined by what 'is'. But it does not follow that 'ought' has nothing to do with 'is'. Hume's experimental method, his focus on human 'passions', and his claim that 'reason is the slave of the passions' form a very different starting point from the abstract notion of individuals' utilities, and Rawls's placing of these abstracted individuals in a state of nature, behind a veil of ignorance about their place in society. In Rawls – and indeed Ramsey and Stern – it is *rational* choice, based upon individual utility, that counts. In Hume, it is human nature embedded in history and society that provides the basis for his theory of justice.[16]

This ethical debate is not just an academic one. It has profound implications for the design of climate change agreements and policy. Stern is right about the importance of ethics. Climate change policy has to be informed by ethical judgements. But those ethics have to be *grounded* in resource allocation, demand and supply, and mediated through actual rather than ideal behaviour – in other words, in economics, although not *determined* by economics. Curiously, Stern's review of the *economics* of climate change fails to make this connection, caught up as it is in Ramsey's *ethics* of a zero time-preference rate.

Carbon consumption

The approach to assigning responsibility based on historical emissions and on fairness creates *moral* arguments for assigning responsibility. To these can be added a third powerful *factual* argument, with its own moral implications – the role of carbon consumption. The importance of grounding this assignment of responsibility on economic facts is that it allows us to focus not only on who caused emissions in the past generally, but also on who is causing emissions now. It is these new – marginal – emissions that we can do something about, and hence take into account for practical efficiency. Efficiency dictates that the focus of attention should be on who and what causes these marginal emissions.

Kyoto was designed to address this question and it came up with the wrong answer. Under the Kyoto Protocol, countries are given caps on their carbon production. A capped country is required to limit how much pollution it emits from its power stations, factories, transport systems and homes. The Kyoto-capped countries – primarily in Europe – accepted such caps, and their emissions have fallen.

This seems an intuitively good – and indeed obvious – way of going about mitigation: reduce the pollution emitted. But there is a snag: what if emissions are reduced in the capped countries, but not in the uncapped ones? And furthermore, what if the capped countries happen to be deindustrializing anyway – swapping domestic industrial production for imports from the uncapped countries?

Let us take an example – Britain. It is a pertinent one, because its performance against its Kyoto production targets is very impressive. Between 1990 and 2005, Britain reduced its carbon emissions by over 15%. It is cruising towards its Kyoto target. This is a record its political leaders are very proud of, and has been used as evidence that Britain is showing world leadership in tackling climate change.[17]

It seems too good to be true. Carbon production appears to have fallen sharply without much by way of economic consequences. Indeed, it *is* too good to be true. For what was taking place between 1990 and 2005 was two processes, both of which are very relevant to our overall argument. Britain was deindustrializing, and it was switching from coal to gas in electricity generation. Neither was a direct result of climate change policy. They would have happened anyway.

The deindustrialization process started at the beginning of the 1980s. A severe recession set in after the oil shock in 1979, which took real oil prices higher than at virtually any time since. Britain lost 25% of its manufacturing output in just 18 months, and the emergence of North Sea oil drove up the exchange rate – in turn squeezing manufacturing further. Britain moved out of making things like steel and chemicals, and into things like professional services and banking.

As a result, the energy ratio – the amount of energy needed to produce a unit of output – cracked. Whereas in the post-war period up to around 1979 economic growth of around 2–3% per annum required an increase in electricity supply of around 7% per annum, now the relationship broke down. More growth did not require lots more energy because that growth was now in services rather than energy-intensive goods. Thus deindustrialization drove down energy demand relative to the post-war trend, and that in turn meant that emissions were ameliorated.

Added to this benign context, as natural gas became legal to burn in power stations after 1990, the share of coal was pushed down and the gas share rose correspondingly in electricity generation. This in turn bore down on emissions – although this switch had little to do with carbon and climate change policies, since there were no meaningful carbon prices for the relevant period. To the extent that policy played a role, it was concern about acid rain which drove up the costs of coal, by requiring the retrofitting of flue-gas desulphurization (FGD) equipment to scrub out the sulphur.

Whilst the coal-to-gas substitution provided a mitigation benefit regardless of what any other country did, the deindustrialization process represented a relocation of energy-intensive industries rather than a global reduction in emissions. For whilst Britain in our example once produced things like steel and chemicals, it now imported them instead – from countries like China. Producing steel in China in less efficient factories, using electricity generated in less efficient coal-fired power stations, and then incurring the emissions costs of shipping the steel across the oceans is, from a carbon perspective, very bad news.

The result is very different from what British politicians present. Adding back those imported emissions caused by manufacturing the products that were previously made in Britain leads to a dramatically altered perspective. Between 1990 and 2005, whilst carbon production fell by 15%, *carbon consumption went up by over 19%*.[18] In other words, the British were *causing* an increase in global emissions, masked by the reduction in the amount produced in Britain. It is this carbon consumption that matters in terms of global agreements and national burdens.

This helps to explain how not just Britain but Europe more generally could reduce its emissions and meet its Kyoto targets, and at the same time have little or no impact on global emissions and hence climate change. Kyoto just made Europe look good and created the illusion of action on *global* climate change. Unfortunately, it is turning out to be an expensive illusion, as Europe drives up its energy costs to achieve its targets. Raising energy prices in Europe if others do nothing just makes matters worse – speeding the deindustrializing process, by reducing competitiveness, *without* reducing carbon consumption.

Europe and the US made matters even worse from a climate change perspective by encouraging excessive consumption (and hence import demand) through the loose monetary policy after the stock market crashes in 2000. This led to a rapid growth in debt-financed consumption in developed countries, notably the US and Britain, which spilled into global trade. This in turn boosted exports from China, which in turn boosted the demand for oil, gas, coal and minerals, sparking a commodities boom. The commodity boom led to greater exploration and production in a host of resource-rich countries, including Australia, Indonesia and Canada. These countries' emissions rose as a consequence. World trade therefore expanded beyond its (carbon-) efficient level.[19]

The Great Recession that followed when the debt-fuelled bubble burst after 2006 had its carbon impact when the US and Europe reduced demand for China's carbon-intensive exports. But its effects were short term, and within a very short period emissions growth returned. This was due to a combination of the stimuli applied by many governments, the now more self-driven growth of a number of Asian and other rapidly developing countries, and domestic demand in China.

If what matters is our carbon consumption, the next question is what causes carbon consumption to rise or fall. There are two obvious answers: income, and the proportion of energy produced by fossil fuels. Income is conventionally measured by GDP on a country basis and GDP per head. The responsibility of consumers in the developed world is measured by the carbon weighting in their consumption. In principle, we could try to calculate this, create a ranking of responsibility, and then allocate the costs of reducing emissions at the global level. In practice, this would be immensely complex, and anyway there is a shorter cut.

If the carbon emissions embedded in goods and services are priced (regardless of where they are produced), then those who are responsible for consuming more carbon through the goods and services they buy will pay more, and those who consume less will correspondingly pay less. To the extent that the goods consumed are made from low-carbon sources, the cost will also be correspondingly less. Carbon pricing therefore addresses both the demand for carbon embedded in goods and services, and the relative economic attractiveness of different ways of generating the energy used in producing them.

The trick is to get that price established. The challenge is not to get the price exactly right, but to make progress away from a situation where carbon is not priced at all and therefore is exactly wrong. Taxing carbon consumption directly is inevitably complicated, and trying to work out the carbon composition of all the goods and services in the economy – including the imports – is practically impossible. What we need is a short cut – something which goes in the right direction, bearing down on the carbon embedded in goods and services, but without all the information and detail required on a case-by-case basis. Fortunately there is a way to cut through all this detail by focusing on the key energy-intensive industries. This is addressed in Chapter 9.

Ascribing responsibility for global emissions is anything but straight-forward. Because ethics are inevitably involved, there is no right answer. We have looked at this from three distinct perspectives: historical respon-sibility for the stock of carbon in the atmosphere; individuals' emissions and their entitlements; and carbon consumption. On all bases, the current approach in the UNFCCC and Kyoto frameworks is seriously misleading. Europe and the US put the carbon up in the atmosphere during and after the Industrial Revolution. Europeans and especially Americans account for most current carbon production per head; and Europeans and the Americans do most of the consuming of carbon-based products.

We can argue about the relative importance that should be attached to each of the three perspectives of past emissions, fairness and carbon consumption. Part will be factual, and much will be moral. But such debates need not stand in the way of making progress. Ethics need to be grounded in (but not determined by) human nature and the circumstances that pervade today, rather than in some future utopia. Progress on this level requires a rejection of the Kyoto-based carbon-production approach and a serious shifting of the burden towards the developed countries. Since all three perspectives point in the same direction, the fact-based carbon-consumption approach, through the establishment of a price for carbon consumed, is an obvious way forward. So why hasn't it happened?

PART TWO

Why is so little being achieved?

CHAPTER 4

Current renewables technologies to the rescue?

How should governments respond to the rising emissions? Not surprisingly, there are lots of ideas, and many lobbies, interest groups and NGOs pushing particular 'solutions' and even 'silver bullets'. Everyone, it seems, has a view, especially when there are potentially large subsidies on offer.

In Europe, many interested parties – and many politicians – are pretty sure they 'know' the answers. They develop 'road maps', scenarios and forecasts, and use these to justify the answers towards which they feel well disposed. Analysis becomes a tool to support predetermined technology choices, and the task of policy design is to effect the rapid deployment of these preferred solutions. It is not just that they are 'picking winners': they are certain that only they can suffice, and the market might 'pick losers' – technologies that they 'know' are bad. Under the technology-driven approach, the government decides which technologies will deliver the outcomes.

In Europe the chosen 'winners' have been the *current* renewables – wind (onshore and offshore), solar (rooftop solar panels and some larger-scale solar installations), and bio-energy (biomass and biofuels) – as opposed to *future* renewables. The confusion of the two is often deliberate: the lobbyists for specific current technologies want to tarnish those who question the economics of these technologies as being against renewables in general. Yet try as they might to protect the subsidies and profits of vested interests, they cannot escape some particularly worrying facts: together, these 'winners' turn out to be some of the most expensive ways known to man to marginally reduce carbon emissions. They are largely ineffective, and in the case of bio-energy they face serious sustainability problems. Here we concentrate on wind, whilst not neglecting the other renewables.

Why wind is so expensive

Working out just why wind is so expensive and has little hope of making much difference to climate change is far from straightforward, and requires a careful navigation through the minefield of misrepresentation that pervades the subject. Whenever a new wind farm is opened, a press release will invariably claim that it will produce enough electricity to power a given number of homes – neglecting to point out that this is only when the wind blows. Press releases are often strangely silent about the costs, but when these do come up, too often they leave out the subsidies and supports, and take a cavalier approach to the system costs of transmission and distribution. Typically too, the back-up costs of conventional technologies needed to provide power when the wind either does not blow or blows too hard are neglected. The press releases of the wind lobby tend to be regurgitated by the media with few caveats and little questioning.

The calculations are not made any easier by the location-specific cost characteristics.[1] A wind farm in the Hebrides islands off the coast of the far north-west of Scotland is very different from one in the south east of England. Scotland has excess supplies of electricity. From a remote location, the electricity generated needs to be transmitted under the sea, then beyond the mountains of western Scotland, on to Glasgow, and finally south into England. Almost any electricity generated in such locations from almost any energy source would be uneconomic except for local consumption. In contrast, a small wind farm connected directly to the London area grid would have negligible transmission and distribution costs, and could supply a market otherwise dependent on supplies from the Midlands, the north, and coastal nuclear stations.

Though each location has, in an important sense, unique elements, there are nevertheless generic features which handicap wind compared with most technologies, whether renewable or otherwise. These are its intermittency; its impact on the security of supply; the consequences of its zero marginal costs of generation; the fact that the best wind locations are very often in more remote, open and valued landscapes; and the sheer amount of land or seafloor it needs to produce electricity. All of these are on top of the costs of the technology itself – the construction of the masts and the turbines, and their maintenance.

The intermittency problem

The first economic handicap for wind is that it is intermittent: the wind varies in strength, and that variance is uncertain. To generate at full capacity, the wind needs to be not too strong, not too weak, and not too gusty. When it is – which is most of the time – something else has to help meet the demand for electricity.

Unlike most commodities, including coal, gas and oil, electricity is not easily stored. Demand has to be instantaneously matched by supply, and, in the absence of large-scale hydro,[2] the storage options that are typically available, such as pumped storage, are expensive. As its name implies, pumped storage involves pumping water uphill at times of excess electricity supply, and letting it fall back down through hydro-turbines when there is excess demand. So far other options such as batteries and heat stores are of little general use, and therefore electricity systems build in extra capacity to meet demand surges and supply failures, and system operators try to persuade some customers to have their supply interrupted to reduce demand at times of system stress.

In the future, all this might change, and the storage of electricity may be a problem that gets solved. There are lots of opportunities. Smart technology may improve the way demand responds too. Some of these promising options for *future* renewables and storage technologies are explored in Chapter 11. But for wind built now, and for the next decade or so, it is a reasonable assumption that intermittency will continue to require lots of back-up rather than storage. This introduces a qualitative difference with almost all fossil fuel technologies, and with many other renewables.

The lobby group Renewables UK claims that the load factor for wind (the time it actually generates relative to its maximum potential) is in the range 25–40%, with an average of around 30%. Since this number really matters for the economics, it is in its interest to present such an optimistic estimate. Other estimates take a much more pessimistic view, suggesting that 20% is a more accurate calculation. The official load factor for onshore wind in the UK was 21.7% in 2010.[3] At these sorts of levels, this is a massive handicap and a major reason why the costs are so high.

To consider the implications, imagine an electricity system where the total demand can be met entirely by wind, when the wind blows. Suppose too that the area covered is one that from time to time lies within a single high-pressure weather system – the sort that northern Europe is prone to, especially in mid-to-late winter. These are characterized by little or no wind. In theory, in addition to the thousands of turbines and supporting infrastructure, there would also need to be a complete non-wind system as well to cover these periods – two sets of electricity capacity, rather than one.

In practice, a 100% no-wind situation is extremely unlikely for a large-scale wind-generation system (although 90% no wind occurs surprisingly often in the UK long-run data), and interconnectors and large-scale, long-distance grids can help to connect across weather systems. But unfortunately this only reduces the size of the problem: it does not eliminate it. The greater the wind-generating capacity relative to the market size, the worse the problem. Thus the next time you hear a politician – like the Scottish First Minister Alex Salmond – claiming that 'by generating all of Scotland's own electricity needs from renewables by 2020, we will be well-placed to go further and become a leading exporter', you will know it is nonsense.[4]

No calculation of the costs of wind can legitimately ignore these additional costs, which are the direct consequence of the scale of the inter- mittency that most other technologies do not suffer from. Yet they almost always do. The obvious question is then: how much *extra* cost does a marginal, additional wind farm add in terms of the back-up requirement? This does not have a clear and unambiguous answer – and for a very important reason. The extra capacity needed as back-up is a *system* cost, not a marginal cost. It cannot easily be disaggregated. So the intermittency costs often depend not on the costs of the specific wind farm, but rather the costs of a system within which wind is a component.

This system cost creates a lot of wiggle room for wind and its lobbyists. For the first few wind farms on the system, the marginal costs imposed are negligible. But this is no longer the case once the total wind capacity crosses a threshold. So, some wind on a system not designed with wind in mind is best thought of in marginal cost terms. It adds very little in the way of extra system costs. But once the threshold is crossed, the system average costs become important. That is the difference between what is on occasion and in favourable locations a good idea – some wind generation – and what is an altogether much more costly idea – a lot of wind.[5] The bad news is that the EU's targets will take all the main European countries over this threshold – of about 20% wind in total capacity.

If we were starting from scratch and intended the sorts of wind genera- tion volumes now being contemplated in Germany, for example, in several of the scenarios the European Commission has set out in its EU 2050 Roadmap,[6] and indeed in Britain, the obvious implication would be that what is needed is a *new system* – designed with this intention in mind. Sadly we are condemned to reality, not some ideal; and, notwithstanding the costs of replacing our networks with this ideal, even a new system would not eliminate these back-up costs.

The extent of this infrastructure cost is only beginning to be under- stood. Wind farms tend to be small scale at locations remote from demand, requiring new transmission lines and the ability of networks to handle big swings in power flows as the wind blows. It is a major industrial undertaking on a scale that has, in many electricity systems, not been seen for more than half a century. Attempting to pinpoint the costs is in a sense to put the cart before the horse.

Suppose for a moment Europe was to go for an energy system based on wind for a significant amount of its electricity. It would require a rewriting of the transmission and distribution map of Europe, including an offshore transmission system, and lots of interconnections to bring hydro back-up to the core industrial and population areas of demand. This is anything but incremental.[7]

Wind lobbyists, faced with these enormous system costs, think they have an ace up their sleeves: all this infrastructure investment will, they claim, create jobs and economic growth. As European Commissioner for Climate Action, Connie Hedegaard, states:

> the added investment would stimulate new sources of growth, preserve existing jobs and create new ones. There is the potential to add 1.5 million new jobs in net terms by 2020.[8]

But this too is far from obvious: the costs would have to be paid, and these would raise prices above those of our competitors blessed with cheap energy, especially gas. Lots of traditional 'brown' jobs in industries sensitive to energy costs would be lost, and in the meantime alternative investments, such as gas infrastructure and liquefied natural gas (LNG) plants, would not materialize. Although it would be nice to pretend the costs don't matter, sadly resources are limited. Hedegaard's claims should be treated as political advocacy, not hard economics. Intermittency is expensive, full stop.

Security costs

The second handicap for wind is that, contrary to the claims of its advocates, it can reduce security of supply and hence raise the system costs. Wind lobbyists claim that it is a more secure source of supply because it is independent of fossil fuel supply and price risks. Wind, it is argued, does not cause price volatility, and is not subject to political interference. Advocates point to interruptions in Europe's gas supplies through Ukraine in 2006 and 2009 (and further difficulties in early 2012), to the Iranian problems and the Strait of Hormuz in the Arabian Gulf in 2011–12, and the trend in oil and gas prices since 2000. Connie Hedegaard claims that:

> the low-carbon transition will need considerable additional investment but our analysis shows this would be largely offset, or even overcompensated, by major reductions in the EU's oil and gas imports.[9]

She further claims that such a transition will reduce 'our vulnerability to potential future oil price shocks'.

This argument is flawed on several levels. First, wind is not a predictable source of electricity. As noted above, its load factor is around 20–30% at best, so for a lot of the time it is not such a reliable supply and, in addition, maintenance problems lead to outages that can last for long periods, especially offshore in winter. It therefore needs fossil fuels to back it up.

It is this last condition that gives rise to an important *additional* and typically ignored security problem. With lots of wind, the fossil fuel supplies required are also rendered intermittent. The gas (or even nuclear) generator cannot rely on security of demand. It therefore cannot contract for fuel supplies on a take-or-pay basis; producer incentives are weakened as gas usually costs money to store. Gas field economics typically point to continuous production. So more wind now means greater gas insecurity, and more gas price volatility as a consequence.

This consideration is compounded by the seasonal *timing* of wind generation. The point at which gas prices are most volatile is when the electricity system demand is high. This tends to be in cold winter weather in Europe. The cold weather is normally caused by high-pressure, stationary air systems over the European continent. These are usually very still, and hence precisely when demand for electricity is at its highest, wind supply can be at its lowest. In these circumstances, wind reduces security at precisely the peak demand points when it is most important.

All these claims ultimately turn on the more general idea that Europe's oil and gas supplies are insecure, that the prices will inevitably go up, and that oil is even relevant in the electricity context. Why? What is the evidence? As we shall see, the irony is that this certainty about the future – that politicians know what the future for oil and gas prices looks like – has been played up at just the point where the fossil fuel supplies have been revolutionized by new technologies. Gas – the fuel that matters – turns out to be anything but scarce.

The marginal cost problem

The third problem with wind is that, once it is deployed on a significant scale, its peculiar cost structure has profound effects on the rest of the generators on an electricity system, driving up their costs. The reason is that wind turbines have *zero marginal cost*, and therefore they will always run and generate when the wind blows favourably. There is no fuel cost – the wind itself is at least free. This means that, at the margin, wind always forces other generators with non-zero marginal costs off the system.

Consider the problem faced by investors in a new gas-fired power station. Investment appraisal starts with their predictions about how much of the time the new power station will be generating electricity, and then the hours it runs are multiplied by the expected price of electricity in the market. Typically, a new station will aim to be baseload – i.e., to run all the time (except for maintenance). But if there is now lots of wind, there will be times when wind generation is the only technology on the system, forcing everything else off, and driving the marginal cost of energy to zero. The implications are radical: it makes the new gas investment much more risky, and it makes gas contracting difficult, since how much gas the power station buys as its fuel depends on factors outside its control: the wind speed and the amount of wind generation on the system.

As the amount of wind on the system grows, a paradox arises. The gas power stations become more important in providing back-up for the wind when it does not blow, but the very fact that the wind generation is on the system means that the costs (and risks) of building the necessary back-up supply rise. This is an additional cost of gas generation *caused* by an increase of wind on the system. In the absence of policy intervention, it leads to two effects: too little back-up will be built, and the costs of what is available will go up.

So serious is this problem in countries with lots of wind (or potentially lots of wind) that their electricity markets are having to be redesigned to create extra incentives to build the back-up, especially for gas-fired power stations. Most European countries are now undergoing significant electricity market reform. Some form of fixed payment, called a capacity payment, is needed as a direct consequence of adding wind to the system, and it is therefore a reflection of the additional cost of wind.

Landscape and other environmental damage

The fourth problem is the costs to the environment. The best locations for onshore wind tend to be upland areas, which are typically environmentally and scenically sensitive. New transmission lines need to be built and, in densely populated Europe, planning issues come to the fore for the siting of both wind farms and transmission lines.

The planning objections are often dismissed as 'nimbyism' – literally 'not in my backyard' – and an unwillingness to embrace landscape degradation as a price worth paying for dealing with climate change. Ed Miliband, the last British Secretary of State for Energy and Climate Change in the 1997–2010 Labour government (and now leader of the Labour Party), suggested that objections to the siting of wind farms and transmission lines should be treated as socially irresponsible in the same way that smoking in pubs and restaurants once was. His successor as Secretary of State in the 2010 Conservative/Liberal Democrat coalition, Chris Huhne, described wind turbines as 'beautiful', and hence as an enhancement to the landscape.[10] Others take a very different view – about both the impact of the wind turbines and the transmission lines. They regard the issue as one of the industrialization of the remaining wilder areas of open landscapes.

Proposals to erect wind turbines on Haworth Moor in the Yorkshire countryside, the inspiration for Emily Brontë's classic nineteenth-century novel *Wuthering Heights*, and to scar the open landscapes of the Scottish glens, have caused outrage. Rich in wildlife, heritage and history, people value these tranquil and largely untouched landscapes. Once development takes place, whatever the developers assert, the damage is likely to be permanent. The environmental impacts are obviously location-specific, and so identifying these, along with the costs, will necessarily be case-specific. Some landscapes have already been so degraded that adding wind turbines will not make much difference. But it is obviously not a good argument for damaging the rest. At one stage there were plans for 10,000 onshore turbines in the British countryside, and it is not hard to see that these would transform the landscape. Caring about this is more than mere nimbyism.

There are also direct environmental impacts. In addition to the obvious concerns about bird strikes (which can be significant), there are the impacts of the footprint on the landscape. In upland areas the soil cover is typically thin. To erect wind turbines requires access and roads. Once built, the roads tend to get used – and not only for the wind farm operators. Moorlands with peat (a very important carbon store in addition to its biodiversity value) are particularly vulnerable to these impacts. Fly over any wind farms in such areas and the patchwork of roads and tracks is all too apparent. Noise is also a consideration – both generally in the open landscape and for those unfortunate enough to have a wind farm next door.

A question of scale

The landscape impact needs to be put into a context: the sheer scale of the land area required to deliver significant electricity supplies. We are not talking about a few wind turbines here and there. To make a difference to emissions, their deployment would have to be on a massive scale – far beyond the 10,000 once mooted for Britain. Before becoming Chief Scientific Adviser at the British Department of Energy and Climate Change, David MacKay did some back-of-the-envelope calculations.[11] If the windiest 10% of the country were covered with windmills (roughly half the area of Wales), he estimates that we might produce the same power per person in a day that it takes to drive the average car 25km. His assumptions are generous – and the requirements at the time he was writing (2008) 'amount to 50 times the entire wind hardware of Denmark; 7 times all the wind farms in Germany; and double the entire fleet of all wind turbines in the world'.[12] Let's be clear: this is covering only *half* the average British car driver's power use. It is also about half the energy used in heating and lighting in MacKay's numbers. He concludes that, 'if we want wind power to truly make a difference, the wind farms must cover a very large area.'[13] The obvious fact is that this is practically – and politically – impossible.

For densely populated European countries, onshore wind can therefore make at best only a marginal difference. Only in countries with large-scale open spaces – and fewer planning restrictions – might onshore wind make a significant impact if all the other problems could be overcome. That means parts of the US, Russia and China, rather than Europe, the main focus of the promotion of wind generation.

The extra problems of offshore wind

Offshore wind adds a whole new cost dimension. It is not difficult to work out why. Consider the context: an unstable saltwater environment, remote from the market, requiring underwater cabling to the mainland, and then transmission lines to take the electricity to its customers. Imagine the process compared with onshore wind. The turbine has to be manufac-tured, loaded onto a boat, transported to its location, and bolted to the sea floor. It has to operate in a corrosive (salt) environment. Maintaining it is a technical challenge in its own right. Unlike oil and gas platforms, helicopters cannot be landed on top of the turbine, and boats moored up in the sea are an altogether different challenge to driving a truck up to the site on land. Fixing something 100 feet up or more *and* out at sea is challenging.

No wonder it is expensive. Yet as an immature technology, costs might be expected to fall. So far progress has been disappointing. As a 2010 UK Energy Research Centre report notes:

> cost reductions anticipated in the late 1990s and early 2000s gave way to dramatic increases in the period from 2005–2009. This report finds evidence that cost increases may have peaked, but does not foresee any meaningful reductions in the period to 2015.[14]

Further ahead, significant cost reductions are again predicted and, as we shall see in Chapter 11 on the new technological possibilities, some of these may materialize, but whether they will close the gap with the other technologies remains to be seen.[15] Since much of the kit is made in China, the cost reductions in manufacture lie there rather than in countries like Britain. Even the Germans and the Danish manufacturers have been losing out. Closer to home the cost reductions are more about maintenance, logistics and the supply chain.

Once the extra costs of erecting and maintaining the wind turbine have been added, and the salt-induced depreciation taken into account, there is then the problem of getting the power to shore. New transmission lines are required for an individual wind turbine close to shore. For a conglomeration of wind farms the issues are rather different. They take on a system characteristic. New transmission networks may be required. This system cost cannot be easily ascribed to the costs of a specific wind farm, but ignoring the transmission costs does not make them go away. On the other side of the equation, there is at least one advantage of offshore wind that needs to be taken into account. Planning permission is much easier – that indeed is why they are offshore (given that the costs are so much higher than onshore). Out of sight, out of mind has been the political rationale.

Planning permission may be easier but, as with onshore wind, the sheer scale of the area required to create a significant supply is enormous. David MacKay again provides some ball-park numbers:

> If we take the total coastline of Britain (length: 3000km), and put a strip of turbines 4km wide all the way around, that strip would have an area of 13000km². That is the area we must fill with turbines to deliver 16 kWh/d [kilowatt-hours/day] per person.[16]

To put the numbers in context, this is again (as with his onshore calculation) less than half the power needed for the average driver per day (40 kWh/d). As a major contribution to reducing emissions, offshore wind is a non-starter. Incredible though it may seem, there just isn't enough suitable shallow water.

There is obviously a much greater supply of *deep* offshore locations. Yet deep means expensive, more steel, and they are typically further from the coast and therefore require more transmission capability. If we really need deep offshore wind to offset the coal-burn in China as a viable option now, then the implications for costs and standards of living would be radical – and politically, economically and socially not at all practical.

Adding up the costs

When the costs of wind are fully taken into account, it is not surprising that virtually no significant wind capacity has been added to the European (or US) electricity systems without subsidy, even in the presence of a carbon price. These subsidies are typically large and long term, and the lobbyists keep asking for more. And despite much optimism, there is little prospect of even onshore wind becoming cost-competitive (in the context of a carbon price) for at least a decade.

How big is the cost gap? Though costs vary by location and by system, it is possible to get some ballpark numbers. There are two ways of doing this: the engineering approach, looking at the costs of each physical bit of a wind farm, and adding them up; or the financial approach, asking how much subsidy is being paid to induce investors to build wind farms.

The engineering approach to costs is typically presented as the *levelized cost*. For any generating plant this is the constant real price of power that would equate the net present value of revenue from the plant's output with the net present value of the cost of production.[17] In other words, it is the price that would cover the costs, sufficient to make the investment economically worthwhile.

What matters in comparing the costs of different technologies is *which* costs are included in this calculation. Would onshore or offshore wind be cost-competitive? Lots of people (in many cases with particular interests) have taken a stab at answering this.[18] The way to get to a pro-wind answer is to make the 'right' assumptions. Wind is capital-intensive, so it is very sensitive to the cost of capital. Assume a low number. Wind is intermittent, so choose a high load factor. For good measure assume that the costs of the turbines will fall – a lot. Then ignore the wider costs listed above. For the comparison, assume that the cost of gas will go up – a lot – and then ignore cost reductions in making and building gas power stations. Finally assume a high carbon price.

If you do all of the above, then the answer that conveniently emerges is that in Britain by 2020 onshore wind just about makes the cut, and offshore wind does not. If, on the other hand, you push back on just some of these assumptions, it is not hard to make onshore wind at least a third more expensive in 2020 than the costs of current electricity generation, and offshore wind anything between two and four times as expensive. Spurious precision does not change the fundamental fact: wind does not make the grade. It is quite wrong to assert that onshore wind is reaching what in the jargon is called 'grid parity', and can survive without subsidy. All the other costs do not disappear just because many wind lobbyists choose to ignore them.

The financial approach looks at the subsidies needed to induce investments. Every country has its own subsidy regime, and its own ways of spreading the system costs across all customers for the networks and the back-up generation. In Britain, customers are forced to buy a specified amount of renewables in their energy mix, and this is translated into a renewables obligation. Wind farms get paid in renewables obligation certificates (ROCs) for meeting this requirement – and get paid again through the price of wholesale electricity. Just to complicate matters further, each type of renewables gets a different number of ROCs per unit of electricity (increasing the subsidy by a factor of two for offshore wind).

It is admittedly one of the most expensive renewables support schemes in the world, but it still gives some idea of the costs to customers. Suppose Britain builds 13 GW of offshore wind by 2020 (reduced from the original plan for 33 GW). The cost penalty by 2020 estimated by the Committee on Climate Change, an independent advisory body to the UK government, is around £2.7 billion.[19] When it is remembered that this is 13 GW of *intermittent* electricity, and all the ancillary costs to customers are excluded, the actual cost is much higher – and this is for generation with *only a 30% load factor*.

Solar

Although we have concentrated here on wind, it is, of course, not the only renewables option. Solar has been added to the mix of preferred and potential technologies. But if wind – and especially offshore wind – is an expensive technology, it pales into insignificance compared with the current technologies of small-scale solar electricity generation in northern latitudes. Indeed, so great are the costs of small-scale solar that they currently make almost anything else look cheap – even offshore wind. As the roofs across northern Europe get covered in solar panels, the subsidies to support their very costly economics get added to customer bills generally, and in some countries even to national debts. Typically the better-off invest and get the subsidies, and the poor pay a disproportionate part of the bill.

As with wind, determining why the costs are so high requires consideration of the nature of the technology and – in northern Europe – the intermittency. So far, the scale of solar investments is too small to make much difference to the system costs (although big enough to impact on customers' bills), and therefore for the moment we can set most of these system costs aside. Where intermittency is concerned, solar at least has the advantage that it tends to peak in the middle of the day, thereby coinciding with daily demand peaks. But it is less reliable in winter when demand tends to be higher (and in those countries where the peak demand in winter tends to be later in the day), thus testing the system capacity.

Solar technology comes in a variety of formats. Solar thermal can be used to heat water and, on a large scale, the sun's rays can be concentrated with large mirrors to create steam, which can then drive a conventional power station (concentrated solar power, or CSP). Such large-scale power generation works best in sunny climates with lots of land availability, since it needs high levels of continuous bright sunshine. North Africa is one such obvious location. DESERTEC is perhaps the most ambitious plan to harvest large-scale solar power and sell it to Europe.[20] Selling it means a long-distance high-voltage transmission link, and transmission costs are therefore very significant. California is another potential location.

Solar photovoltaics (PV) convert sunlight directly into electricity. The resulting electricity is used to supply the home, and any surplus can be sold back to the grid. The technology for rooftop insulations is small-scale, and can be retrofitted to most houses. But here the advantages come up against costs and the limits of scale. Solar roof panels have been encouraged across Europe with generous feed-in tariffs (FiTs), notably in Spain, Germany and Britain. The costs are truly enormous – indeed, so much so that the budgetary consequences have led to the subsidies being cut in all these countries, in some cases retrospectively.[21] In the case of Germany, one study calculates the costs of promoting them between 2000 and 2010 at €53 billion.[22] Solar farms can also be built on a much bigger scale but, as with wind, they need a lot of land, and they also industrialize the landscape.

Yet there is good news for solar in the longer term. Whereas the potential for cost reductions in wind generation is clearly limited, solar costs have fallen sharply, and there are good technical reasons for expecting further falls in the future. It has one additional enormous advantage: there is lots of potential energy. One hour of sunshine is very roughly equal to the world's electricity generation for a year. In theory, it is a way of meeting the wall of demand, caused by economic and population growth. Fortunately, as we shall see in Chapter 11, solar remains one of the better medium-to-long-term technologies. This, however, will do little to change the costs of the more inefficient and expensive solar panels now being installed.

Bio-energy: biomass and biofuels

In the scramble to meet European renewables targets, biomass has been seen as a way of utilizing existing electricity generation and heating systems in an apparently low-carbon way. It is likely to provide half of the Renewables Directive requirement by 2020 on current trends.[23] Yet biomass is not all it seems – and not always what its lobbyists claim – and it has become very controversial.[24]

Biomass is considered a renewable in so far as, in theory, it is a closed-carbon loop. Trees and energy crops 'harvest' carbon, which is then released back into the atmosphere through burning in one way or another – in a power station, a car's engine, or a log-burning stove. The biomaterial comes in a host of shapes and sizes: specific crops can be grown for burning or fuel production, and biowaste can be burnt too. Almost any plant material could in theory be used.

Yet the closed loop is less than perfect, and the degree of carbon leakage in this process is important. Just because it is conventionally referred to as renewable for the purposes of the EU Directive does not mean that it actually is renewable. It can be very energy-intensive. Energy is used in harvesting the crop and felling the trees. These 'fuels' then have to be rendered into a state fit to burn – for example, by creating wood pellets. They are bulky, and hence potentially costly to handle, and bulk transport (by ship, road and rail) is involved. This is before the energy loss in the burning process is taken into account. All these steps involve extra energy, and where this is carbon-based, the loop leaks.

Biomass also introduces a timing element. It is a carbon store or sink. It will naturally release its carbon over time, unless buried and stored as peat or other fossil fuel deposits. Some of it will go into the soil and may be captured for centuries. Burning the tree or timber waste puts the carbon immediately into the atmosphere, thereby negating the storage dimensions. In a context in which emissions are rising quickly, it may be more sustainable to leave it as a carbon store until some future date when there are lots of other technologies and we may be reducing the carbon stock in the atmosphere. Biomass is a natural method of carbon capture and storage (CCS). It is somewhat ironic that just as CCS technologies are being developed to address carbon emissions, biomass is being encouraged to do the opposite.

Direct leaks in the carbon loop are not the only problem. Biocrops take up land from other users. Large-scale fields of energy crops like elephant grass have significant agricultural impacts. Marginal land may have to be brought into production – in the process helping to release carbon from the soil. Forests may be deliberately cleared, using the biomass as a by-product.

So far these broader forms of biomass fuel for electricity generation have had limited impact, albeit potentially damaging, yet they pale into insignificance when placed alongside biocrops for the production of biofuels to replace the fuel in motor vehicles and aircraft. Land transport is still largely based around the internal combustion engine, and such engines are dependent on a high-quality direct fuel supply. Oil fits the bill, and the renewables contribution comes through creating biofuels with similar properties for either blending or complete substitution.

Making biofuels is not a new science, and there are a number of bio-sources, many of which have considerable environmental costs, and many involve the release of carbon in their production. Crops require irrigation, fertilizers and energy for sowing and harvesting. They need to be transported to the refinery, which itself uses energy. Then there is the impact of the land-use changes on carbon emissions. Adding all this up can make some biofuels even negative in their carbon impacts – i.e., after production, processing and burning they actually emit more carbon than they mitigate. Perhaps the most controversial is the destruction of rainforests to clear land for biofuel crops – either directly, or by displacing other forms of agriculture, such as relocating cattle ranches closer to the rainforests. Switching land to biofuel production means not using it for some other purpose. If the current use is as a carbon store, the net carbon savings from biofuels will be limited.

The US biofuel programme has been based on the use of corn to produce ethanol.[25] The motivation has in large measure been an attempt to move towards energy independence rather than to address climate change, particularly following the initiatives of the first George W. Bush administration from 2000, as set out in the 2001 US National Energy Policy.[26] All of this is of course pre-shale gas.

Ethanol production has been underpinned by significant subsidies paid to Midwest farmers, who divert corn from the food chain into ethanol production. This in turn raises the price of corn; and for people at the margin of subsistence in the developing world, the price increases have exacerbated malnutrition and hunger.

But like all agricultural subsidies, since being implemented, this one has been fiercely protected by a powerful lobby. There is a 'carbon pork barrel' to lobby for. The higher corn prices raise land prices, and higher land prices are sustained as long as the subsidies flow. Any reduction in corn subsidy means a capital loss to landowners – a further incentive to lobby for the continuation of the policy.

In some countries, biofuels have been produced on such a scale as to make substitution from fossil fuels a significant feature of the economy. Brazil's ethanol production from sugar cane plays this role, although again it is not at all clear that a closed carbon loop has been created. Alternative land use is a key factor, and as Brazil slowly but relentlessly destroys its rainforest, the impacts on one of the world's great carbon sinks enter the equation. In addition, the Amazon's role in regulating climate (including water vapour), and as one of the world's great biodiversity reservoirs, needs to be considered.

As with wind and solar, there is a question of scale. Such an inefficient process of converting crops to fuel requires enormous land areas, and typically lots of water too. In reality there is no such thing as empty or even marginal land – all have their own biodiversity and ecosystems. Brazil and the US have been able to set aside significant areas. Europe has not – and cannot. As David MacKay puts it, the result for Europe at least is that 'biofuels can't add up', and that 'biofuels made from plants in a European country like Britain can deliver so little power, I think they are scarcely worth talking about.'[27] Christopher Knittel has calculated how much US farmland would be required to run America's cars solely on E85 (85% ethanol and 15% gasoline). He finds that:

> if ethanol used corn as a feedstock, this would imply roughly 415 million acres of corn crop – but there is currently only 406 million acres of farmed land in the United States.[28]

Even in those countries less densely populated, biofuels come up against the constraints of agriculture. As global population rises towards 9 billion, and as economic growth drives up food demand and encourages a switch towards meat production, the pressure on land (and water to support agriculture) can only increase. Short of a major technology shift (possibly through genetic engineering), it makes little sense globally to push conventional biofuels. New biofuels, like algae, may have much more potential because they use less land and create useful by-products for conventional electricity generation.

How did Europe end up in this mess?

How could so much money be spent to so little effect? The answer is a complicated one, and combines political motives and ideology with lobbying and rent-seeking by companies in the wind business, support from interested academics and other technical bodies, and a profound misunderstanding of the nature of the fossil fuel markets against which the renewables have to compete.

The instrumental answer to the question of how the money got spent is the EU's Renewables Directive (and the 2008 Climate and Energy Package of which it is a part). It put in place a renewables target for the EU as a whole, with the allocation of specific targets to each country to reflect its circumstances. This target was for energy in general, not just electricity.[29] The EU also mandated 10% biofuels by 2020, and thereby also legally committed its members to a currently expensive and potentially destructive path for transport – with little net climate change benefit.[30]

The overarching Climate and Energy Package was given a catchy title: *2020-20-20*. Everything would add up to the magic number 20 – 20% carbon reductions; 20% renewables; 20% energy efficiency. All in 2020. The idea that the right answers all added up to 20 is laughable – and economically absurd. Climate change is a long-term problem: it has multiple facets, and the 20% renewables target skewed the market towards a small number of what are in most cases very expensive technologies. For some countries, hydro provided a cheap option, but for most it meant wind and biomass (plus renewable heat) and some solar.[31]

The extreme case is Britain. It was given a 15% renewables energy target, on the grounds that its starting point was lower than every other EU member state apart from Cyprus and Malta. Given that there was limited scope outside electricity generation, the 15% energy target translated into a requirement for a very large-scale wind programme within a decade. Given too that there was limited scope onshore due to public opposition, it appeared to imply that as much as 33 GW of offshore wind might even be necessary.[32] Even taking account of the lower electricity demand resulting from the economic crisis, the number was well beyond what most would have thought achievable – and indeed around 10 GW now looks a more realistic aspiration.[33]

The initial projected costs of this crash programme in wind were truly staggering. Early on, £100 billion (not including the full network and back-up costs) was officially suggested for the full 33 GW.[34] Translating that into annual investment, it is £10 billion a year for a decade – just on offshore wind. This would be a significant share of Britain's total capital expenditure. Alarm bells should have been ringing.

Revenue streams are needed to underpin these investments. As there is little coming from the taxpayer, the money has to come from consumers. Few bothered to consider whether they could actually afford to pay. Yet the question is very real: by 2015 perhaps a quarter of all British households may be spending more than 10% of their disposable income on domestic energy, and will therefore be in fuel poverty. Similar numbers may also arise in Germany, and indeed elsewhere across Europe. For industrial customers, it was assumed that the renewables would not be too great a burden, in part because the belief was that fossil fuel prices would rise, and hence the relative cost of renewables would fall. Competitiveness considerations hardly figured at all. No politician considered what might happen to European industry in the event that shale gas transformed its competitors in the US.

The Renewables Directive has had one other paradoxical side effect. As more renewables are forced onto the system, the carbon price in the EU ETS is likely to fall. This increases the competitiveness of coal and gas, which consequently increase their share of electricity generation, so that the resulting emissions reductions are lower and at the limit completely offset. Countries with large-scale renewables programmes will end up selling on their surplus emissions permits, so that others can increase their emissions.[35] By not thinking through the implications of fixing the total quantities of emissions through the EU ETS, and then reducing the quantity through the Renewables Directive, the EU failed to see that the Directive might have no impact on total emissions.[36]

Politics, not economics

It is a remarkable achievement to drive up costs, reduce competitiveness and security of supply, and still make little impact on emissions. To choose one of the most expensive options for carbon mitigation first – ahead of other much cheaper options – cannot be the result purely of faulty analysis. How did the EU choose this path, and enshrine it as legislation through the Directives?

The answer turns out to be largely political. For many green politicians and supporting NGOs in Europe, the real villain of the piece is nuclear. Greenpeace was anti-nuclear (the 'peace' bit) long before it discovered the media impact of seals being clubbed to death, spilling red blood on the white ice of Canada's Arctic.[37] Friends of the Earth was also anti-nuclear in its origins. Many other anti-nuclear movements in Europe developed on the back of the Cold War as the very real threat of nuclear annihilation hung over the densely populated countries of western Europe. (In the less prosperous Soviet east, anti-nuclear activists found it hard to get off the ground.)

This anti-nuclear position is not primarily an economic one. It has deep ideological roots. Closing down nuclear reactors in Europe has been a core goal of the green movement, and in this it has had significant success. But the trouble with closing nuclear now is that it leaves a carbon gap. Greens argue that this should be filled with renewables (and energy efficiency, to which we turn in the next chapter).

Renewables also have the merit of being typically small-scale and decentralized, and this also fits well with the green ideology of sustainable communities, living in harmony with nature. They fit into an 'alternative' way of thinking.[38] Some even see them as beautiful, echoing E.F. Schumacher's famous theory of *Small is Beautiful*.[39] There can be little doubt that Europe's preference for wind has been in large measure driven by the rise of green political parties and green NGOs.

This green preference for renewables is not, however, sufficient to explain how the EU ended up with the 2020 Renewables Directive. For this to be achieved the greens needed to gain political power, and that is precisely what has happened in the last two decades in Europe.

Political power is not just about winning seats in parliaments. It is about influencing the behaviours of all the political actors. As environmental concerns grew amongst the European electorate in the great economic boom of the late twentieth century, all major political parties tried to reach out and capture this emerging political interest. Thus Prime Minister

Margaret Thatcher famously 'discovered' global warming and tried to link concern for the environment with the Conservative idea of stewardship and conserving the past.[40] Parties on the Left were quick to follow, especially where the politics of socialism and green interests coincided (greens tend to be on the Left rather than the Right, in favour of significant income redistribution, as noted in Chapter 3).

But one more ingredient was required: the decline of mass support for the main political parties in almost every European country. By the end of the first decade of the twenty-first century, almost all European countries needed coalitions to govern, including Britain. And whereas the natural coalition party of choice was once some sort of centre liberal party, the electorates of Europe, to varying degrees, replaced the liberals with the greens. The results were messy. In Britain, the Liberal Democratic Party has been, in large measure, a green party. In Germany it was more straightforward, with the greens achieving an electoral breakthrough, and entering directly into government with the SPD (the Social Democratic Party), creating the first major red–green coalition.

Courting the greens inevitably meant ceding control over energy policy. In almost every case of greens entering government, they got the energy ministry – in Germany, Ireland, and in effect through the Liberal Democrats in Britain. In this role, the price has been first an anti-nuclear stance (notably in Germany), and then a large-scale renewables programme.

This is how Europe got its short-term 20% renewable energy target. But it would be too great a compliment to the green political movement to give them all the credit for the Renewables Directive. They have been aided and abetted by industrial interests for whom renewables represent a very large carbon pork barrel. Subsidies attract industry, and with guaranteed contracts and political support, major European companies began to sing the greens' tune. Siemens led the way in Germany.[41] But there were others: DONG in Denmark and a host of manufacturers. The renewables lobby groups grew in size and influence, and the lobbying became overt, loud and very effective, funded by the companies that stand to gain most from the subsidies.

As with all political trends, the support holds up as long as the voters buy into the basic story. The real tragedy of the large-scale dash-for-wind (and to a lesser extent the other current renewables) is that the public have been misled. The costs are not likely to be what were promised, and the public does not always buy into the landscape intrusion wind represents. There is little support for the new transmission lines, and the relentless march of emissions at the global level raises in the public mind the question of what contribution wind is really making to the climate change objective.

One of the ways in which the renewables costs are masked – at least in the eyes of politicians – is through energy efficiency. Not only is it argued that the costs of renewables will not be as great as the assumed rise in fossil fuel prices, but also that energy efficiency measures will reduce the demand for energy to such an extent that the renewables costs will be absorbed in lower total bills. It is argued that the unit price of electricity may be raised by the renewables, but the number of units will fall such that the overall bill will fall. As with the assumption about fossil fuel prices, the claims about energy efficiency turn out to be both more complex than typically presented, and often based on flimsy foundations. In the next chapter we address the energy efficiency arguments, and consider fossil fuel price projections in Chapter 7.

CHAPTER 5

Can demand be cut?

If conventional renewables on the supply side are not going to close the carbon gap – at least until new technologies come along – then what about the demand side? Do we really need more energy? Couldn't we get by using less energy more efficiently? This is the second great hope of governments, green parties and green NGOs.

The conventional argument is that energy demand can be decoupled from economic growth, so we can have more consumption with less energy. We just need to be more efficient in the ways we use energy. This, it is argued, is an open goal, as there are enormous opportunities to increase energy efficiency at zero cost or better. In other words we can increase energy efficiency and cut energy demand *and* make a profit at the same time. This demand reduction, it is further argued, will reduce emissions and, by cutting energy bills, enable customers to absorb the higher costs of renewables. As the British Secretary of State, Chris Huhne, boldly stated in September 2011:

> British energy consumers will on average be better off in 2020 thanks to our low carbon policies. Yes, I said better off. Getting off the oil and gas price hook and onto clean, green energy makes sense. And with energy saving, we can offset the higher prices and end up with lower bills.[1]

We will return to his very questionable assumptions about oil and gas prices later, but even without these it is a very bold prediction about energy demand reductions which he assumes will follow from his energy efficiency measures. It would, if correct, be fantastic – a painless route to decarbonization.

It gets even better. It is argued that by going down the energy efficiency route, we can also get green growth and create lots of green jobs. And there is more: energy efficiency, it is argued, can increase security of supply, improve competitiveness and reduce exposure to the level and volatility of fossil fuel prices. It therefore hits all the buttons: emissions reductions, green growth, jobs, more security and lower bills. With so many claimed merits, it must surely be the closest thing energy policy has to a silver bullet.

This is yet another example of something that seems just too good to be true. As we shall see, energy efficiency has a lot going for it, but it is unfortunately not going to solve the emissions problem, and many of the other claimed benefits turn out to be somewhat less than they seem. To see why, there are four questions that need answering. Does energy efficiency reduce demand? Are there lots of energy efficiency opportunities that are already economic and ripe for the taking? Is global demand likely to fall? And finally, if it is all so attractive and economic, why hasn't policy made much difference?

Does energy efficiency reduce energy demand?

Let's start with the relationship between energy efficiency and energy demand. Energy efficiency is, *by definition*, a good thing: to claim that something is efficient is to claim that it is least-cost. Nobody can be in favour of not minimizing cost. So we want to be as energy-efficient as possible – we can all agree on this. Nothing in this chapter suggests otherwise.

Assume for a moment that there is all this inefficiency about – lots of 'low-hanging fruit', as the energy efficiency lobbyists tell us. Wouldn't this *automatically* translate into a fall in demand? It seems obvious – but it's a big mistake. The crucial point is that an increase in energy efficiency represents a fall in the cost of achieving a particular output – say a warm house. It costs you less as a result – that is the point. Otherwise, why do it? A reduction in cost is equivalent to a reduction in the price of energy.

So what is the impact of a fall in the price of energy on the demand for energy? Economists are obsessed with changes in prices, and they break down the effect into two parts: a substitution effect, and an income effect. The substitution effect of a drop in price *always* leads to an increase in demand. This is the most elementary economics. There is no way around this: the demand curve simply reflects the fact that the less expensive something is, the more it will be purchased. And so it has been in energy for centuries. Jevons famously pointed this out in the nineteenth century, and he got it right (as opposed to his views on peak coal reported in Chapter 2): 'It is wholly a confusion of ideas to suppose that the economical use of fuel is equivalent to a diminished consumption. The very contrary is the truth'.[2]

This is now referred to as the Jevons Paradox. Energy efficiency has improved a lot since his time, in ways he could probably not have imagined. The early steam engines have given way to much more efficient technologies; electricity has gained market share; and oil has replaced coal as the fuel for transport. The modern car and the modern power station are massively more efficient than the horse and cart and steam engines.

As these energy improvements have mounted up, and as energy has become abundant and cheap, so the demand for it has risen. The nineteenth- and twentieth-century industrialization was possible only through the application of vast amounts of non-human energy. Efficiency, price and demand joined in a close embrace. The process continues: every year we get more energy-efficient.

What about the income effect? A fall in the price – because of the increase in energy efficiency – leaves more money in people's pockets. If people spend less money on heating their houses, they have more income left to spend on other things. So what do they spend it on? It turns out that quite a lot of the income saved on energy bills is spent on more energy. This is called the *rebound effect*.

One way it might be spent is directly on more energy for the household, by turning up the thermostat. This is the direct rebound effect, and it is what has been going on for a long time. Since 1970, the average house in Britain has witnessed a temperature rise from around 13°C to 18°C. Where once people sat around in sweaters, they now want to sit around in T-shirts. They demand hot (and hotter) water on tap in copious amounts, with power showers and all the associated services. Today houses are packed with energy-hungry electronics equipment.

There comes a point when houses are hot enough. But that does not exhaust the rebound effect. Looking ahead, the desire for heat may tail off (although there is still an awfully large number of poorly heated houses), but there is a lot more energy that homes could consume. There is air conditioning – almost unknown in the offices and houses of 1970s Britain, but now ubiquitous for offices and creeping into the home. And then there are all those appliances and gadgets that need charging – even cleaning your teeth and reading a book can now require electricity. This is the indirect rebound effect and satiation is likely to be some way off. As is to be expected, there is considerable dispute about the scale of these rebound effects and the way they work through the economy. Yet it is hard to conclude that they are trivial.[3]

So the impact of a reduction in the cost of energy through energy efficiency is unlikely to lead to the sorts of reductions in demand often claimed, taking both the substitution and income effects into account. Since energy efficiency won't do that, neither will it increase energy security much. It is not substituting demand for foreign oil and gas with something else. The important economic fact is that what primarily drives energy demand is the price of energy, and it is *increases* in the price of energy that induce more energy efficiency. Not surprisingly, the big gains in energy efficiency have *followed* price shocks, like those of the 1970s and those of the last decade. For an increase in price increases the returns on investing the time, effort and money in saving energy.

Are there lots of energy efficiency opportunities out there?

Even if energy efficiency is unlikely to reduce energy demand, it would nevertheless be a good idea to take up the opportunities to maximize it. Efficiency is, as noted, an unambiguously good thing. But how inefficient are the major economies? Is there really lots of low-hanging fruit?

Many analysts claim that there is a lot of inefficiency about. The argument tends to focus on buildings, which are responsible for an awful lot of energy use. But it also applies to everything from standby power for computers, to all those lights left on in offices overnight, which make our cities literally glow in the dark. If only houses were insulated properly, appliances were designed to turn themselves off, and people were less lazy and apathetic when it comes to turning off the lights we would use less energy. Then there are opportunities in the industrial and transport sectors. Aggregated over the whole economy, it is argued that these efficiency improvements would amount to a significant fall in energy demand.

The usual starting point in answering this complicated question is to look at energy savings that *could* be made, and then compare these with actual consumption, and argue that the difference represents the *potential* energy efficiency savings. This is called the engineering approach – it is what an engineer would do. Next, the cost of making the improvements is estimated, and as long as it is less than the identified benefits, it is declared amongst the projects with a positive return. Finally, all these profitable projects are added together, giving an aggregate energy efficiency return.

This sort of engineering analysis is reflected in some of the most optimistic representations of the energy efficiency contribution to carbon mitigation. For example, in 2009 the commercial consulting firm, McKinsey & Company, made the remarkable claim that the energy efficiency opportunities for the US economy are 'vast':

the US economy has the potential to reduce annual non-transportation energy consumption by roughly 23 per cent by 2020, eliminating more than $1.2 trillion in waste – well beyond the $520 billion upfront investment (not including the program costs) that would be required. The reduction in energy use would also result in . . . the equivalent of taking the entire US fleet of passenger vehicles and light trucks off the road.[4]

It's the sort of number politicians and efficiency advocates love to quote, but its very size makes it suspect. A cursory glance would lead you to believe that energy efficiency is fantastically cheaper than any other alternative, and that you could triple your money. Could the US economy really be wasting a quarter of its energy? Energy it could save without a net cost – indeed, at a staggering rate of return? In just ten years? And excluding transport – which, if inefficient on a similar scale, would push the overall alleged inefficiency of the US economy to over one third? The short answer is no: it turns out that it is based on very shaky empirical foundations.[5]

This matters, because extraordinary – and implausible – though these numbers are, they are too often faithfully reproduced in government documents. It leads governments to think that they really could make these savings at better than no net cost, and hence to the delusion that climate change can be quite easily cracked. Such numbers find their way into Nicholas Stern's work too. He relies surprisingly heavily on McKinsey's work generally to justify his overall cost estimates for mitigating climate change, notably in *Blueprint*, his follow-up to the Stern Review.[6]

Such enormous returns in any other market would be expected to be snapped up. It is equivalent to walking down the street and seeing dollar bills lying on the pavement and, in the McKinsey case, dollar bills equal to almost a quarter of the US's energy needs. Literally billions. Indeed, it is even better, since many of them are not just any dollar bills, but your own. You could make an incredibly profitable investment in energy efficiency, and yet despite these claims being repeated for decades, the dollar bills stubbornly remain on the pavement and people just walk on by.

Something fishy is going on here, and it is worth exploring in a bit of detail how people could be so stupid as to ignore all the 'free' money, or alternatively what lies in their way. It really matters because it is this dollar-bills-in-the-street argument that is used to justify all sorts of energy efficiency policies and political interventions – and why they rarely deliver what their advocates predict.

It is often asserted that there are 'barriers' to picking the money up. The investments in energy efficiency cost money, and need to be financed. Despite the profitability of the investments, perhaps householders cannot persuade their bank or building society to lend to them. But does this really stack up? Why exactly can't they borrow? Homeowners can borrow to buy their houses against the collateral of the value of the house. The house value should rise to capitalize the returns on the energy efficiency measures, since the house will be cheaper to heat. Homeowners borrow for all sorts of other items, and often these are unsecured loans. In the case of houses, specialist mortgage providers are well placed to understand the returns. It might be more difficult for those renting rather than owning their house. The landlord might not cooperate, being more interested in short-term rents than long-term returns. But again, why won't the extra costs of energy in a poorly insulated house depress the rents? In many cases, the issue here is more about poverty, particularly housing poverty, and the fact that the tenants may not have the money to pay the rent in the first place. It is no accident that poor people live in poor quality – and badly insulated – housing.[7]

Maybe then it is because people don't know the dollars are in the street. They may lack information, not realizing that if they invested in energy efficiency they would be better off. If so, it would not be unique to energy efficiency. There are a whole host of opportunities we don't know about until someone tells us, and indeed one of the ways the market works is that businesses have an incentive to tell us in order to sell their products. Governments too can tell us, and there is no shortage of advertisements and public information when it comes to energy efficiency.

It could even be that we really are too stupid to pick the money up. Perhaps we cannot do the sums and are incapable of acting in a fully rational way. There is now a burgeoning behavioural literature on the latter – limited or bounded rationality.[8] The problem with the behavioural approach is not only that it is hard to think of a good reason why it would be especially relevant to energy efficiency – as opposed to all our other big choices – but also that it is too easy to start with the *assumption* that the returns are positive, observe that projects are not taken up, and put the gap down to individuals' limitations. In other words, assume the conclusion and iterate backwards. A further step easily slips in – that since the government *knows* the answers, psychologically limited people need to be guided – or even forced – to the 'right' answer.

Psychological failings are not necessarily a good reason for concluding that the returns are positive. For it is one thing to say that people are limited in their rational appraisals, and quite another to say that the government is any better. The consumer lives in his or her house; the bureaucrat lives somewhere else. There is almost no discussion of 'government failures' to set alongside the claimed market failures. As we shall see below, programmes to enhance energy efficiency typically *assume* that there are no government failures. The bureaucrat is *assumed* to know best, despite being at an informational disadvantage in the context-specific circumstances.

To be fair, this problem is sometimes recognized, and leads to the search for credible institutions whose incentives are not so obviously subject to this bias. Trusts, mutuals and NGOs are often utilized for these purposes, but again they often have their own agenda – especially NGOs, whose own budgets depend on there being 'a problem' to solve. Institutions have objectives, and too often the temptation is to make the evidence serve these objectives. There is a large and vociferous energy efficiency lobby to match and complement the renewables lobby.

There is just one other argument and it is one that the energy efficiency lobby find hard to swallow, for it questions their very rationale. The dollar bills may not exist. McKinsey may have been wrong – indeed, very wrong. Consider the analytical approach outlined above, and the position of the person trying to decide whether to make an energy efficiency investment. Few have got hours to spare doing a detailed analysis. Although there are many in fuel poverty, there are many who are not, and for most, the energy bill is just another utility bill. If the price of energy shoots up, then that gets their attention. There is a price effect. When oil and gas prices are low, energy efficiency does not matter so much.

In order to sort out which investments are really efficient we need to go beyond the crude engineering studies and start by defining what we mean by 'efficient'. This is *net*, not gross, of the full costs of the investment, and these costs are direct and indirect. A householder needs to sort out first whether there are gains to be had. Should the government and company adverts be taken at face value? Or should they be treated as if a health warning applies? This is the first cost – trying to sort out spin from fact. The next step is to work out what would actually be involved in making the investment. This means getting builders and fitters into the house. Can you trust them? Will they turn up on time? Will they complete the job? Will it involve you taking time off work? Next: will they do a good job? Will they clean up afterwards? Will there be additional damage to make good? There is likely to be lots of hassle involved and lots of additional cost factors. Many simply can't face it.

These considerations lead to a rather different conclusion: it is far from obvious that there are lots of economical energy efficiency measures out there ripe for the picking, even if there are lots of engineering ones. We could *technically* make lots of savings, but we could similarly technically do all sorts of things. Just because they are technically possible, it does not make them economic.

Another way of thinking about this is to consider the typical market reaction to a gold mine like this. The obvious solution is for energy-savings companies to emerge which are paid from the energy savings themselves. Given the huge scale of the opportunities claimed, the economic rents should attract major new entrants, and companies should be able to exploit these opportunities. If McKinsey was correct about the enormous scope for energy efficiency, these energy-savings companies would be in the top ranks of the main share indexes – up there with Microsoft, Apple and Google. Almost a quarter of US demand would be an enormous market. But they are not. They hardly exist at all. What should we conclude? That companies – like householders – are too stupid to pick up the dollar bills lying in the street? That they have not read this report, and they have not taken the consultancy's advice? Or perhaps the companies cannot substantiate these apparent economic rents?

The focus so far has been on household energy efficiency to illustrate some of the practical issues. It is in part motivated by the fact that our homes are responsible for lots of emissions; that householders face the main barriers to take-up; and because household energy efficiency is meant to cut bills and hence make the renewables costs more bearable. This latter point has a political dimension: consumers ultimately have to pay for climate change mitigation, and hence it matters what their bills are.

But it is also important to remember that governments and business all use a lot of energy. When it comes to governments, it is not hard to think that there may well be many energy efficiency opportunities that are not taken up. Government incentives are complex; budgets are compartmentalized; and there is no direct link between cost savings, career prospects and departmental interests. It is an area where there is much work to be done. Indeed it may well turn out that governments waste more energy than households.

Looking beyond buildings and governments, transport is not only a major area of energy use, but also one to which energy efficiency can – and has been – applied. The improvements have been considerable. Miles per gallon keep going up. But as with the buildings debate, it is far from clear whether energy efficiency drives the demand for energy for transport, or the price drives the incentives to improve vehicle efficiency. It is no accident that heavily taxed motorists in Europe drive smaller cars that do more miles per gallon compared with their American counterparts. Efficiency standards have played a role in this, as have car companies' research departments, but it is not obvious whether they have made as much difference as the price.

This brings us to one final concern about whether the level of energy efficiency is high enough: whether the price reflects the full costs of the energy used, and particularly the associated emissions. For if it does not, energy efficiency will be sub-optimal. Putting a price on pollution – and particularly on carbon, as discussed in Part Three – increases the returns to energy efficiency. Given that energy efficiency is a good thing, and given that we might want the polluters to pay for their emissions, householders, businesses, governments and drivers should be incentivized to make the investments which pass the economic test at the full price of energy.

What is likely to happen to global energy demand?

Increasing the price of energy changes the economic returns, and rational consumers react by demanding less. In recent decades there have been two periods when the price of energy has risen substantially – following the oil shocks of the 1970s, and particularly the shock in 1978/79, and after 2000. In both cases, the increase in energy prices preceded a drop in economic growth,[9] and hence there was a negative income effect (their spending power was reduced) and a reduction in the trend growth of demand because consumers expected the price increases to be permanent. The price rises got their attention. The early 1980s was a period of recession in many developed countries, and the big price increases in the mid-to-late 2000s were accompanied by a global recession.

The demand response in the 1980s and beyond was highly significant in two respects: not only did the relationship between economic growth and energy demand change, but the composition of developed countries' economies started to change too. In Britain, for example, a major structural shift away from energy-intensive industries got under way, changing energy demand patterns permanently. In the period since the Second World War up to the end of the 1970s, in contrast, a 3% growth in GDP meant a 7% increase in the demand for electricity. The graph opposite illustrates the scale of this fall in the British energy ratio, which was already on a downward path.

The breakdown of the traditional relationship between economic growth and energy demand in Britain has been repeated to a greater or

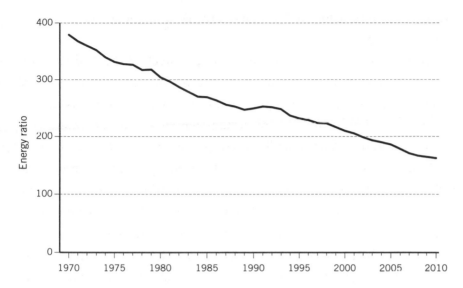

Energy ratio in Britain 1970–2010 (tonnes of oil equivalent per £1 million GDP, 2006 prices)

Source: Department of Energy and Climate Change, *Digest of United Kingdom Energy Statistics 2011*, London: The Stationery Office, 2011, Table 1.1.4

lesser extent across developed countries.[10] It is a reflection not just of increased energy efficiency but also of the deindustrialization process. But the picture in the developing world has been less encouraging. Just as Europe deindustrialized, losing its energy-intensive industries, so China and other developing countries industrialized and increased the role of their energy-intensive industries (even though they may have been becoming more energy-efficient too). This is really just another way of reflecting the key point in Chapter 3 that whilst Europe reduced its carbon production, it increased its carbon consumption by importing these energy-intensive goods from the likes of China that it once would have produced itself. The falling energy ratio reflects the fact that countries like Britain no longer have much industry in the energy-intensive sectors, as well as the increases in the energy efficiency of what is left. Steel, chemicals, aluminium and so on now come from abroad. Whilst the energy ratios have been falling in Europe (and elsewhere), at the global level energy demand has continued on an upward path with GDP growth, with developing countries' demand growth swamping the developed country trends.

This is a pattern that has some way to go. Much of the medium-term growth in energy demand will come from the increased incomes of the poor and near-poor in developing countries.[11] This transition is energy-intensive, and as households make the step from poverty towards the middle class, they acquire appliances (especially refrigerators) and cars.

What this means is that the growth of energy demand through to 2050 – projected by the US Energy Information Administration (EIA) and the IEA, for example – may underestimate the wall of demand coming from the rising incomes and populations in the developing world: the extra 2–3 billion with economic growth rates of between 5% and 10% per annum. Whilst the developed world might be able to hold demand very steady, and attempt to gradually decarbonize the supply needed to meet it, developing countries face a much greater uphill struggle.

Two factors might mitigate this trend. The first is price: if energy prices rise, demand falls. Thus, if the recently experienced high oil and gas prices were to be extended, and if the forecasts of $200/barrel for oil were to be realized, then demand would be much lower. Such a price effect would also encourage investment in energy efficiency measures, and one might expect the falls of the energy ratio to be further embedded in the structures of the economy. This, in turn, would induce greater R&D in the efficient use of energy (such as new materials, smart technologies, more efficient boilers and new ways of capturing and preserving energy and heat in buildings). It might also encourage 'catch-up', as developing countries like China copy these investments in order to maintain competitiveness in a high-energy-price world.

However, the price effect factor suffers from one major problem: what if fossil fuel prices do not rise? What if fossil fuels turn out to be abundant and cheap? This is what happened against expectations in the 1980s. The conventional wisdom at the time was that the oil price spike in 1978/79 had changed the game, and that it would turn out to be a floor and not a ceiling for future prices – a bit like the assumption about $100/barrel for oil in the last couple of years. In fact the opposite happened. The good news is that the energy efficiency investments were practically irreversible. The bad news is that the diffusion of these improvements – the catch-up – in many developing countries may be a very long process, and more immediately lower prices of fossil fuels would undermine the incentives.

So what might the future demand for energy look like? What is the likely trend around which energy efficiency measures and price changes might revolve? There are lots of forecasts, and all require a series of assumptions – about economic growth, the price, and the elasticity of demand (the responsiveness of demand to a change in price). All, too, implicitly make assumptions about technology – that it doesn't change much. The graph below presents recent IEA forecasts.

What the graph tells us is that there needs to be a step-change in price and technology. Existing measures will make only a limited dent in the trend. The troubling fact is that for all the energy efficiency measures, energy demand at the global level is going to keep going up. It matters little

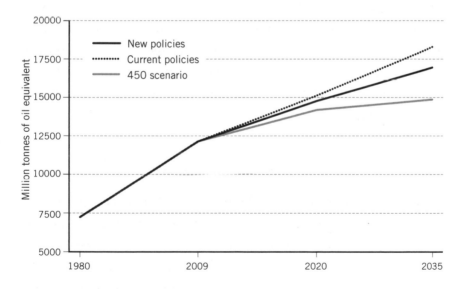

IEA forecast of global energy demand to 2035 for three policy scenarios (million tonnes of oil equivalent)

Source: International Energy Agency, World Energy Outlook 2011

whether, whatever the specific merits, all the houses are insulated in a developed country like Britain. What matter are all those extra cars, fridges and heating and cooling systems in India, China and other developing countries. Under current policies, the IEA's scenario to 2035 has global energy demand increasing by over 50%, and by 40% under 'new policies', under which recent policies are implemented in a cautious way. This is the climate change problem on the demand side, driven by economic growth and population – to complement the increasing fossil fuel-burn on the supply side. It is the macro problem, to which very micro European solutions have been proposed.

A carbon price would help dampen energy demand as well as increase energy efficiency, as discussed earlier. There is just one other option – not to have the economic growth in the first place. Some more fundamentalist greens suggest we can get by without more consumption, as well as less energy. They may be right, and it may be that the wider consequences of all the economic growth that may be coming our way, and the population growth, are just not compatible with sustaining the wider environment. This is altogether more radical and, whatever the ethical merits, shows no sign of gaining any serious political traction. If the answer to climate change is no growth, we are not going to crack the problem. Fortunately, as explained in Part Three, there are good prospects – with more enlightened policies – of us getting by with some economic growth and limiting climate change.

Energy efficiency policies

The wall of demand does not undermine the validity of energy efficiency measures. It simply undermines the idea that energy efficiency can, with renewables, achieve the overarching objective of halting the global increase in emissions. Its contribution is valuable, and if there is a lot of technical change it might be very valuable, as discussed in Part Three. But it is unlikely to be decisive. How then have energy efficiency measures worked out so far? How have at least three decades of initiatives in developed countries changed the energy efficiency of these economies (separate from the effects of energy prices)?

There are several approaches to energy efficiency policy, and they are all incredibly 'micro' in focus. The overwhelming emphasis has been on buildings and the domestic sector. As already noted, part of this is a desire to be seen to be doing something about energy costs for voters and fuel poverty. But there has also been a steady pressure to drive up energy efficiency standards, and this has been applied to cars, appliances and to building regulations.

Let's start with the domestic housing. The approach is to identify particular individuals and particular housing stock, and make targeted direct investments in them. The obvious candidates are those in fuel poverty and social housing, yielding not only a benefit for the recipients but also reducing the claims on wider public expenditure by addressing the fuel poverty.

These programmes are often carried out by housing associations and local government, and are in practice extensions of housing policy. Where the state is the owner of the housing, it is acting in its landlord role. Energy utilities are involved to an increasing extent, with regulation obliging the utilities to implement energy efficiency programmes and raise levies from the customer base through the energy bills. Evidence on the success of these programmes is mixed for a number of reasons: the impact of the general price rise on bills increases the returns to energy efficiency, as distinct from the direct measures carried out; poorer customers tend to have larger rebound effects; and because the counterfactual – what would have happened in the absence of the intervention – is rarely specified.[12]

A second option is to try to address the claimed financial barriers by easing the credit provision. Here governments can either provide loans and grants, or underwrite the loans. In the latter case, efficiency measures can be tied to the finance of the house and its meter, thereby reducing the cost of capital, and getting round the problem of people moving house.

A third option is to provide information – to tell people that the dollar bills are lying in the street. Often the information required is the result of specialist knowledge, and governments can provide 'energy audits' and 'energy ratings' to buildings and appliances. Most governments do in fact do this in developed countries. Again it is important to understand what this information includes – and what it leaves out. It is typically 'engineering' information – about the physical characteristics and energy usage – and excludes the economic analysis of costs. Thus its effect may be limited, but desirable where there are economies of scale in providing the information and, crucially, when it is credible.

Many governments promote new packages or deals. These tend to start with the politics – announcing some ambitious target to treat all the housing in a country by a certain date (and typically claiming that thousands of jobs will be created in the process), or some overall target for achieving energy efficiency improvements. Given that public funds are scarce, the deals tend to focus on private revenue streams. These can come from either the direct beneficiary of the measures, or all energy consumers. The former translates into a finance package; the latter into some form of levy. Most government policies pursue both.

An example is the Green Deal in Britain, which offers energy efficiency surveys, identifying the potential gains for householders, and then financing the work from the householder's (increased) electricity bills.[13] It therefore targets informational failures and capital market failures. Again in the British case, a levy is also applied to all consumer bills to raise funds for energy efficiency measures that do not pass the positive returns test (often called the 'golden rule'), notably for households in fuel poverty as part of a broader social policy. All consumers are in effect taxed for the benefit of poorer consumers in badly insulated homes.

The success of these policies depends on the willingness of households to participate; the ability of the providers to raise the finance; and the willingness of consumers in general to pay a tax on their energy bills for the benefit of those receiving the levy-backed measures. But nobody should be seduced into assuming that as a result demand will fall.

Let us now turn to standards, and for the purposes of illustration concentrate on transport (although they also apply to buildings, and in particular, building regulations for new homes). The standards approach is a command-and-control-type of regulation: the government decides the 'right' answer, and then imposes these conditions. In the command-and-control tradition, the approach has been to define what the 'best available technology' is, and require this as the standard. More recently, the approach has been to apply standards beyond the technologies available, to force technical progress.

The Corporate Average Fuel Economy (CAFE) legislation for car efficiency standards in the US is a contentious example.[14] It has been much debated as to whether CAFE made much difference compared with what manufacturers were doing anyway, and the extent to which its impacts can be disentangled from the effects of changes in fuel prices. Motorists in the US – and their politicians – are notoriously unwilling to tax fuel. Successive presidents from Nixon to Carter, Clinton and Obama have failed to raise taxes. When CAFE was introduced, sports utility vehicles (SUVs) were conveniently defined as light trucks, and hence had a much lower standard applied to them. The results were reflected in a boom in SUV sales in the US. The EU has imposed its own fuel efficiency standards, the impacts of which (as in the US) are much debated. As with energy efficiency generally, the rebound effect is an important component of any analysis of the net effect.

Standards have a further dimension in their international impact. Countries exporting to markets that have such standards in place have to comply with them. This is a way of enforcing international standards indirectly – a bottom-up approach. In principle this helps to get round the problem of international coordination and agreements. Countries can unilaterally impose standards in their own markets, thereby forcing their consumers to pay the price of the pollution abatement mandated by the standards, regardless of whether the products are made at home or imported.

The problem with standards is deciding what they should be, and this is one of the reasons why economists prefer market-based interventions and especially prices. How much pollution should be allowed? Should cost come into the calculation, given that higher standards typically mean higher costs? How would government know? The reality is that standards are a paradise for lobbyists, who try to gain traction on the politicians. In the US, for example, the car industry saw off any tightening of CAFE standards for three decades.

How big is the energy efficiency prize?

The energy efficiency argument is based on grand claims about the scale of the prize, with little evidence that the gains are anything like as big as suggested. Nobody can object to making energy consumption as efficient as possible. The question is: how big are the potential gains with positive returns? Contrary to what many politicians claim, the answer is not obvious. The *claimed* gains are enormous, but the response of individuals and companies is *small*. *Economic* energy efficiency may be very different from technical or engineering energy efficiency.

No one can doubt that governments have tried hard to solve this problem. There have been energy efficiency measures advanced across developed (and some developing) countries since at least the 1970s. These policies have had mixed results, and indeed in many cases what has been called energy efficiency measures are really social policy (with a correspondingly large rebound effect).

The energy efficiency role in climate change policy is based on two flawed arguments: that energy efficiency will substantially reduce general energy demand; and that there are lots of projects with positive returns. The former is hard to substantiate – and indeed historical experience has proved otherwise.

The latter – the 'low-hanging fruit' argument – is open to serious doubt, and depends in any event on the price of energy. It is here that yet another mistake is made in the conventional arguments about both energy efficiency and renewables: that fossil fuel prices can be *assumed* to be going ever upwards, so that both energy efficiency measures and renewables will become more economic. This is far from certain. On the contrary, as we will see in Chapter 7, we are probably moving into an age of even more fossil fuel abundance in the areas that matter – coal and gas – even if (and it is an 'if', not a 'when') oil prices stay up or rise further.

The implication is that whatever the merits of energy efficiency policies, they probably will not make much difference to the global emissions path, short of a technological breakthrough. The good news is that this just might eventually be possible, as we reveal in Part Three.

CHAPTER 6

A new dawn for nuclear?

Alongside renewables and energy efficiency, another technology has been championed by some as a solution to climate change. Nuclear power advocates promise zero-carbon energy, and they claim that it is a competitive and secure source of supply. The climate change agenda has, some claim, given nuclear a new lease of life after three decades in the doldrums. They argue that only it can provide large-scale, low-carbon electricity generation.

The green NGOs, by contrast, have a long history of opposing nuclear, both military and civil. The reasons are deep and profound. Their strategy of current renewables plus energy efficiency could be expressed another way: as the alternative to nuclear. Indeed, so strong is the 'anything-but-nuclear' commitment that some German greens in particular might even prefer rising emissions to nuclear power.

There is nothing new about the claim that nuclear power is economic, and cost-competitive with fossil fuels. From the outset, some even said that it would be 'too cheap to metre'. But it wasn't, and the low fossil fuel prices in the 1980s and 1990s, coupled with the partial meltdown at Three Mile Island in 1979 and the explosion at Chernobyl in 1986, put paid to most development in these decades. The years of the great expansion of nuclear power were in the decade 1965 to 1975. After 1980, beyond

France's borders, new nuclear build was a rare phenomenon, particularly in the US.

Climate change was heaven-sent for the nuclear industry. Even if it could not compete directly with fossil fuels, it had an enormous advantage – it could provide baseload large-scale electricity generation that was apparently low-carbon. Suddenly its traditional rival, coal, was no longer a long-term option, and its competitor was now presented not as coal, but rather renewables. Add in a carbon price, and allow it to compete head on with renewables, and there suddenly appeared the possibility of a new dawn. It might not have been able to compete with coal, but the high costs of wind and solar made it look much more attractive. Nuclear, having been on the environmental defensive for decades, now portrayed itself as the green option. Climate change was seen by nuclear advocates as a game-changer.

The reality has been rather different. As the oil price rose, nuclear was gradually clawing itself back onto the agenda in the early years of this century, and a number of countries began to get serious about building new nuclear plants. In 2007–08, 24 new reactors were proposed in the US. Then along came two major setbacks.

The first emerged gradually: the discovery and exploitation of shale gas, and the recognition that gas might be abundant and cheap for some time to come. This pulled the economic rug from under nuclear's feet: it had been relying on peak oil and ever-rising oil prices, which in turn would drive up coal and gas prices.

The second was Fukushima. The series of meltdowns and explosions at the nuclear plant following the catastrophic tsunami which hit Japan in March 2011 provided graphic media images around the world, and provoked panic in many electorates and in some politicians. If even the Japanese could not work out that if they built a nuclear reactor near a geological fault line on the coast – and if a tsunami came along, it might knock out the back-up power systems and hence trigger a meltdown – then how could anyone feel safe?

Gas and Fukushima may have been major setbacks for nuclear, yet neither has proved a knockout blow (yet), and nuclear remains one of the options for large-scale electricity generation. But it is unlikely to fill that role any time soon, and in order to see why nuclear in the climate change context is no more a silver bullet than current wind, current solar, or energy efficiency, we need to take a good hard look at nuclear's past, and its underlying economics.

A brief history of nuclear

Civil nuclear power grew out of the military programme at the end of the Second World War. Many civil programmes are run by countries with nuclear weapons, and in the US, Britain and France, the civil programme initially supported the military, in particular through the production of plutonium. This relationship remains in many countries an intimate one. Even in Germany and Japan, and a number of other European countries that had no nuclear weapons themselves, these nuclear industries existed in a context in which they got their technologies from the US and shared in the nuclear umbrella that the US provided to Europe and Japan (supported too by British and French nuclear weapons). It is therefore not surprising that those who oppose nuclear weapons also tend to be very suspicious of civil programmes.

During the 1950s and 1960s, the US began to develop its civil programme on the back of rate-of-return regulation, passing through the costs to consumers. In Britain, and later France, state-owned enterprises developed nuclear programmes, and spread the costs across consumers too. Statutory monopolies meant that these costs could be imposed on them. There were no *purely* commercial reactors – all relied on the state in one way or another. That generally remains the case today[1] – especially where liberalization and competition have replaced monopoly, and consumers can therefore switch suppliers.

The immediate nuclear challenge for Britain after the Second World War was to build its own independent bomb to maintain its 'great power' status. All the United Nations Security Council permanent members went down this route. The problem for Britain was making enough of its own plutonium, and the early reactors at what is now Sellafield (formerly Windscale) had plutonium production as a major driver.

The reactor design chosen in Britain was one where this priority so dominated that little thought was given to the resulting waste. The MAGNOX graphite reactor was a waste-intensive technology, and this feature fed across to the advanced gas-cooled reactors (AGRs) that followed. By 1990, the then head of the Central Electricity Generating Board (CEGB), Walter Marshall, could claim that the British reactors had produced more waste than all the rest of the world's nuclear industries combined.[2] It was a problem that had not been thought through at the outset, and one that has dogged the industry ever since. No country has yet come up with a particularly convincing solution, although some (notably Sweden) have made more progress than others.

The global nuclear industry prospered throughout the 1960s and 1970s. By 1980 there were around 400 reactors around the world, with over 100 in the US alone. Arab oil producers, through their revitalized oil cartel OPEC, caused a major oil shock in the early 1970s, and this reinforced a widespread conventional wisdom amongst the developed economies that one solution to the exposure to OPEC had to be nuclear. When the Iranian Revolution came along at the end of the 1970s, and the oil price rose to $39 per barrel (higher than at almost any time since in real terms), the serious doubts that had begun to surface about nuclear's costs were less convincing. The energy-intensive industries in the developed economies needed more and more electricity generation, and nuclear was regarded by many as providing security of supply.

France was the most dramatic in its commitment to a nuclear programme. It had had the traumatic experience of the Algerian War; it had, along with Britain, failed at Suez; and it had few natural resources of its own. Energy independence became a driving political objective and nuclear fitted the bill. France therefore embarked on a programme of new pressurized water reactors (PWRs), favouring US technology over French alternatives after a fierce internal battle, and eventually brought 59 single-design reactors onto its electricity system, producing around 80% of its electricity. It is the most nuclear country in the world.

Others planned to follow the French lead, notably Britain under Margaret Thatcher. Believing the oil price could only ever go up, a plan to build ten PWRs was announced in 1979, at the rate of one per year. Japan, as another energy-resource-poor economy, also took the path to a major nuclear programme. By the time of the Fukushima accident in 2011, 51 reactors were in operation as the result of a continuous new-build programme from the 1970s. And there were plans for many more (although six months after Fukushima, Japan began a review of its nuclear energy policy). Yet not everyone delivered: in Britain, 14 years after the new-build decision, only one of the planned reactors was actually built (Sizewell B), and privatization of the nuclear electricity industry in the 1990s ground new-build ambitions to a halt. The impact of falling oil prices, and the legalization of gas as a power station fuel after 1990 in Europe, undermined the economics of new nuclear, which was now exposed to the harsh scrutiny of the private financial markets. In the US, the partial meltdown at Three Mile Island produced a backlash, and no new nuclear power stations were constructed thereafter. Canada carried on for a bit longer. However, it would take rising oil and gas prices after 2000 to rekindle interest internationally.

The nuclear industry now looked to the Asian markets for its future and, prior to Fukushima, the growth of China, India and South Korea offered new opportunities, as the chart opposite indicates.

The green NGOs and nuclear

In political and policy circles, it is very hard to have a pragmatic and rational debate about nuclear power. Green NGOs regard nuclear as a dangerous technology that should be stopped in its tracks. This is an ideological position for most; a matter of faith rather than a question open to debate and discussion. As a result, they tend to treat anyone who is not

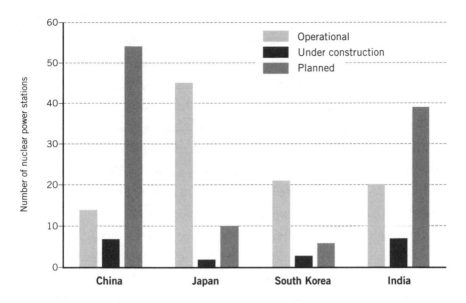

Number of nuclear power stations in China, Japan, South Korea and India, 2012

Note: Japan's nuclear policy is currently under review, and as of May 2012 all of its nuclear plants were offline.

Source: World Nuclear Association Information pages, www.world-nuclear.org/info/inf63.html

against nuclear as pro-nuclear, and often fail to appreciate that it is possible to be agnostic – neither ideologically for, nor ideologically against.

In Europe, anti-nuclear protests began in the 1950s and grew in the 1960s. The threat of nuclear annihilation was very real on the front line of the Cold War, and it was a natural step from opposing nuclear weapons to opposing civil nuclear power. Indeed, the two were, as noted above, linked. The Campaign for Nuclear Disarmament (CND) had a long history in Britain, with its annual Aldermaston March, and later Germany witnessed the protests at the deployment of cruise missiles (which Helmut Schmidt had been so keen on for fear of the US weakening in its support for a western Germany) and at the Gorleben waste disposal site. In the US, anti-war protests focused on Vietnam in the late 1960s, but it did not experience the broad anti-nuclear movement on a scale comparable to that in Europe.

The green NGOs made little progress until the turn of the century in their campaign against civil nuclear power. They did not need to outside

France, as the economics undermined the case for new-build. From 1980, new nuclear was not cost-competitive. But in the last decade, with a certain sense of irony, as the climate change argument intensified the green NGOs gained some success against nuclear. The most spectacular case came in Germany as the Green Party emerged as a major electoral force. In 2000, the new Chancellor, Gerhard Schröder, formed a coalition between his Social Democratic Party and the Green Party. The reward of coalition for the latter was a gradual nuclear exit. Yet the green parties' electoral success did not immediately deliver a non-nuclear future, even in Germany. That would take another nuclear accident.

In the last decade a number of European countries – along with Japan, China, India (see the chart above), the US and Russia – looked to nuclear as a way out of the threat to energy security, higher oil prices (as they had in the 1970s), and to address climate change. In Britain, Prime Minister Tony Blair repeatedly tried to get a new nuclear build programme going, but was thwarted initially by senior Cabinet colleagues with CND backgrounds. Indeed, the 2003 White Paper on energy policy, Britain's answer to how it would decarbonize, all but ruled out nuclear in favour of a renewables plus energy efficiency strategy.[3] It was not until 2005 (after his third election victory) that Blair could move on to a positive position on new build, to be set out in a White Paper in 2008 (after he had resigned).[4] This more pro-nuclear policy was forced through into the Coalition agreement after the 2010 election, with even the incoming Liberal Democrat energy minister, Chris Huhne, agreeing that nuclear was essential in the decarbonization effort (having described it in opposition as 'too expensive, too costly, and we shouldn't go down that road'[5]). However, it would not prove easy to translate ambition into concrete action.[6]

In Germany, after the defeat of the Red–Green government led by Schröder, Angela Merkel gradually edged her centre-right coalition towards tentative pro-nuclear steps. This included lifting the deadline on closing the existing nuclear power stations, through to the possibility of life extensions, in return for a nuclear fuel rods tax. Her government was, however, still a very long way from new build, and indeed German industry (notably Siemens) had backed out of the nuclear cooperation with the French (and Areva).

But within a few weeks of announcing the policy on life extensions, Fukushima happened, and the German government did a spectacular U-turn. Now not only would Germany close all nuclear stations, but eight of them would be immediately shut down. So German nuclear policy had gone from a phase-out, to a life extension plus a tax, to an immediate closure of some and complete closure of all within a decade – and a fuel rod tax to boot. Neighbours were not consulted, and the consequences were immediately felt across the European energy markets. As if to reinforce the *energi wende* ('energy turning') as it became known in Germany, the Green Party beat the SPD into third place in the Baden-Württemberg regional elections shortly after the Fukushima nuclear accident, and in coalition the regional parliament got the first Green Party minister-president in Germany.

The immediate closure meant that more coal and gas would be burnt in Germany in the short and medium term (and Germany would need to import more nuclear as well as coal- and gas-generated electricity from its neighbours, whilst it built up its renewables, now on an even more ambitious timetable). Carbon emissions would be much higher. When asked about his views on this shortly after the German announcement, the Polish Prime Minister said that the decision to shut down the nuclear plants was 'great news' for coal-intensive Poland: it could now export more coal to Germany, and more coal-fired power stations would be built:

> From the Polish point of view it is a good, not bad, decision, since we are fighting to continue to be able to use Polish coal. If one of the largest countries in Europe gives up on nuclear, then what source of energy will do the trick? Windmills, with all due respect for green energy, will not fill the gap which the closure of over 20 nuclear power plants in Germany has created.[7]

For the greens, climate change was a difficult issue, notwithstanding the long-standing hatred of nuclear. There were nagging doubts. Was climate change so serious a threat that even nuclear had to be contemplated? There were a few defectors within the green ranks, the most high-profile being Patrick Moore, one of the founders of Greenpeace.[8] But typically the doubts came after they had left.

But strip away the ideology and there were a series of issues raised by nuclear, which deserved serious analysis and which left many in the agnostic camp outside the NGOs with cause for concern. It is not necessary to be in a green NGO to have doubts about the role that nuclear can – or should – play in climate change mitigation. The four main concerns are: the very small risk of a catastrophic accident; the waste; weapons proliferation; and the carbon emissions from the nuclear cycle. Each needs careful analysis and each has implications for the economics.

The very small risk of a catastrophe

Under the conventional approach to risk assessment, the probability of an event is multiplied by the expected damage caused by that event. Applied to nuclear, the probability of a meltdown with serious associated leaks is extremely small. In three such events – Three Mile Island, Chernobyl and Fukushima – the damage caused has not been 'catastrophic', although it has undeniably been bad. In the Three Mile Island case, nobody died, and there was no large-scale escape of radiation. In the Chernobyl case, people died. Radiation spread across Europe, and many have been adversely affected. There is much controversy about whether and to what extent it resulted in additional cancer deaths.[9] There were significant economic losses to farming and the food industry. In the Fukushima case, it is likely that there will be several deaths as a result of the heroic acts of workers who tried to cool the reactors; there will be deaths from cancer among the wider population; and there will be economic damage in the surrounding areas for years to come.[10]

All of this is bad, but in discussing nuclear events, it is very difficult for people to keep some perspective, especially given the Armageddon stories in the media and the activities of green NGOs. But let's consider these cases compared with coal mining. As we discussed in Chapter 2, anyone who works in a coal mine is likely to damage their health to some extent. Indeed, if the same standards were applied to coal mines as to nuclear power stations, including in respect of radiation and related health impacts, it is hard to think of many coal mines that would not be closed down. Each year thousands die in coal-mining accidents, including the 5000 per annum in China referred to previously. Beyond the specific cases noted above, the nuclear power industry has probably not directly killed a thousand in its entire history, and most of the deaths that have occurred will have been due to conventional accidents. In Japan, many thousands died because of the failure of the authorities to protect the population from the tsunami, but very few died directly because of the nuclear explosions and meltdowns at Fukushima. The cases of cancer that are yet to emerge will have to be compared with the lung disease of miners.

Multiplying the very low probabilities by the damage will necessarily produce a very low number. But many argue that this sort of calculation should not be used when considering catastrophic risk. Empirical evidence suggests that we place more weight on such events than indicated by the probabilities. But what does 'catastrophic' mean in the nuclear case, and should this principle of extra weight be applied elsewhere, or is it specific to nuclear?

What would a catastrophic nuclear accident look like? Fukushima was not the end of the world, or even of Japan. The reactors have melted down, there has been some radiation leakage, and people have been killed and many exposed to radiation. More will die from cancer in due course.[11] But is this *catastrophic*? Or is it a very serious accident with containable and measurable consequences? It could of course have been worse, but again what is the worst-case scenario? Could it have destroyed a whole city? Japan? Asia?

These risks should not of course be downplayed, but it is a reality that in many aspects of life we run large-scale risks with potentially even worse consequences. We allow antibiotics to be used indiscriminately, risking the development of resistant bacteria which could kill millions. We allow dangerous dictators to remain in office and commit genocide – Mao's regime may have killed as many as 70 million Chinese. The Iranian leadership might launch a nuclear attack. We take inadequate precaution against tsunamis, which, as we saw in Japan, can annihilate entire communities.[12]

If we are to put more weight on small probabilities of large-scale events, then how do these stack up against deaths that happen routinely and about which we do little. Are lots of people killed and injured in one big bang more important than the daily carnage on the roads? The British government 'allows' around 2500 people to be killed on the roads *each year*, and around 5000 to be seriously injured. This could be radically reduced by imposing a speed limit of 20 miles per hour everywhere, but, not surprisingly, this is not done.

Nuclear risk matters, but it is not so great as to rule the technology out; if we did so, a whole host of other candidates would be waiting in the wings, including coal mining and cars. The risks need to be taken into account, and they need to be protected against. The consequence shows up in the economics of nuclear, which we deal with below. These costs of mitigating the risks may well turn out to render nuclear uneconomic, but that is entirely different from arguing that the risk rules out nuclear *per se*.

The waste

Nuclear waste is fundamentally different from other types of waste. Some of it is very dangerous and takes centuries to decay. Plutonium as a waste product from nuclear power stations has a half-life of around 24000 years. On almost any ethical basis, it is beholden upon those who create such waste to come up with a solution for dealing with it, and the principle stated in the Flowers Report way back in 1976 is a good one to work with:

> There should be no commitment to a large programme of nuclear fission power until it has been demonstrated beyond reasonable doubt that a method exists to ensure the safe containment of long-lived, highly radio-active waste for the indefinite future.[13]

Yet the early history of the nuclear industry reflected an almost total disregard for the waste issues. In Britain, the first official White Paper on the subject in 1959 suggested that there were plenty of old quarries and disused mine shafts that could be used for disposal.[14] On the military side, dumping at sea and the abandonment of nuclear submarines was not unheard of. None of this was helped by the secrecy that surrounded the nuclear industry, and it comes as little surprise that nuclear opponents have little confidence in much of what the industry says.

So what are the options for dealing with the waste? These boil down to three: bury it in a deep depository; store it above ground; and reprocess it. The first takes the long half-lives as a given, and seeks to place it in geological structures that are radiation-proof, and beyond access and use. The second is a wait-and-see strategy, which allows for new technologies to emerge. The third option is to partially recycle – for example, by creating mixed oxide fuel (MOx) to be burnt in existing reactors – or the waste could potentially be consumed in closed cycles that use up the plutonium in part or in whole.

It is easy to understand both why the deep depository option is the preferred route for all the main civil nuclear industries, and why it gives rise to concern. The timescales are so long that there can be no certainty about political stability, and little confidence that the waste could be 'safe' for such periods. On the other hand, if future generations want to get their hands on plutonium, there are lots of ways they can do this – just as we can now – and hence nuclear waste is not a unique issue when it comes to the risks from plutonium.

The second option, above-ground storage, is the cheapest in the short term: it avoids – or delays – the costs of building a deep depository. It also has the advantage of adjustment and flexibility if new technologies do come along. This is where the third option comes in. It is not at all clear that plutonium is in fact 'waste'. It could be a valuable fuel, used to generate more electricity. The dream of the nuclear industry has been the fast breeder reactor, which would consume the waste in a closed cycle. Current reprocessing plants re-use some of the spent fuel, but they still leave a lot of waste. New designs offer more radical opportunities targeted at the plutonium, and in theory are capable of reducing the half-life of the plutonium to a much more manageable 300 years, and using the resultant fuel to generate electricity. If deliverable, such possibilities could close off quite a lot of the nuclear cycle.[15]

This matters a lot. If the nuclear cycle could be at least largely closed off, then nuclear as an option for addressing climate change becomes a whole lot more sustainable. Then there would be the option of making big carbon reductions without the small scale and intermittency that renewables bring. But it is an 'if' not a 'when', and even then there are still issues to address.

Weapons proliferation

Nuclear weapons do create catastrophic risk. A nuclear war could destroy whole areas of the planet and kill millions. After the Second World War, and especially during the Cold War, many lived in fear of this happening, and it is not surprising that many demonstrated and protested against nuclear weapons.

The threat has not gone away. The secretive activities of countries such as South Korea and the prospect of Iranian nuclear weapons raise considerable international concern. Yet, so far, after more than 60 years, no one has exploded a nuclear bomb in conflict since Hiroshima and Nagasaki. This, as Thomas Schelling has remarked, is one of the most surprising turns of events in the period. Despite the proliferation, despite the rogue states with nuclear capability, nuclear weapons have not been used.[16]

The link with civil nuclear power has been a very real one, and the creation of plutonium that can be used in nuclear weapons is an obvious by-product (and in some cases more than a by-product). But should civil nuclear power be abandoned because it produces plutonium that could be used in nuclear weapons? The fact is that nuclear weapons are not going to go away, and military nuclear powers are not short of plutonium. The worry is that new countries will obtain nuclear weapons, and that terrorists might do as well. Building more civil nuclear power stations in Europe, the US and China will not make much difference to the availability of plutonium to those who seek to become nuclear powers. Indeed, if the plutonium can be used to produce electricity in new reactors, this raises the possibility that the stockpile of nuclear weapons could be rendered beyond use and turned into electricity.

Carbon emissions from the nuclear cycle

The risks of a catastrophic accident, the waste and proliferation are serious, but arguably not show-stoppers. They leave nuclear with some serious problems and costs, but it remains a candidate to consider in addressing climate change – provided that nuclear does in fact reduce carbon emissions. Some have called the carbon credentials of nuclear into question. A new nuclear power station embeds a lot of carbon, and it takes a lot of energy to build it. Then there is the uranium mining, another energy-intensive activity (with additional environmental problems associated with it). Nuclear is not, therefore, carbon-free. But then there *are* no carbon-free technologies. Carbon is embedded in wind farms and solar panels. These also take a lot of energy to manufacture and construct, and in the case of offshore wind, the ships use diesel. So it is a question of *relative* carbon emissions. If that carbon is priced, this will wash through into the economics; if not, we need to include the carbon production in the mining and plant manufacture and compare it with the other options over the lifetime of the plant. Here again it matters a lot whether the plutonium is re-used and the cycle closed, or whether continuous uranium mining is called for.

The economics of new nuclear

Nuclear is different from other technologies in that a great deal of weight is placed on its specific characteristics, and especially the risk, waste and proliferation issues. But it also has a number of economic characteristics that have limited its contribution to electricity generation over the last few decades, and that continue to dog the technology going forward.

A nuclear power station is essentially a very large lump of capital, which takes five years (sometimes more) to build in Europe or the US (less in China). It then has very low marginal costs and operates for decades, before facing large decommissioning costs at the end. It also has to provide for handling the waste, although the profile of costs depends on whether a 'wait-and-see' strategy is deployed (and hence the costs of dealing with it are postponed), or whether it is dealt with immediately.

New nuclear power stations have all the usual sorts of costs and risks associated with such large-scale construction projects, and there are also lots of safety costs to reduce risk – much more so than with any other electricity-generating technology.

These characteristics mean that a key variable in the economics of a new nuclear power station is the cost of capital. The higher the cost of capital, the less competitive it is against its main conventional competitor, the combined-cycle gas turbine (CCGT) with its mix of fuel costs and rapid construction. This is a feature that nuclear generation shares with wind farms and many solar installations.

Construction costs have historically tended to rise, rather than fall, with experience. The adverse trend began back in the 1970s in the US, and this experience surprisingly extends even to the French PWR single-design programme.[17] The latest French reactors, Olkiluoto in Finland and Flamanville in France, are both late and significantly over budget.

A second variable is the forecast cost of gas or coal that fuels the competing conventional power stations. If the gas price is expected to soar then nuclear is much more economic. If it falls, then it is priced out of the market. In the late 1970s, oil and coal prices (gas was not used for power stations then, or indeed ever expected to be used) were predicted to go ever upwards, and hence utilities in the US, France and Britain proposed big new-build programmes. The subsequent collapse of fossil fuel prices wiped out the economic case for nuclear for the next two decades. In the early years of this century, 1970s-style forecasts were being proclaimed again. The arrival of shale gas changed all that in the US, and the forecasts of future gas prices undercut nuclear, even if the carbon price is significant.[18]

The importance of the capital costs and the exposure to fossil fuel prices make investing in nuclear a very difficult proposition for the private sector. Thus, investors seek to strike a bargain with consumers: if the nuclear power station is built for their benefit, then they must commit to pay back the capital costs. In other words, there needs to be a long-term contract. The same applies to wind developments, especially offshore. But since future consumers cannot commit now to this sort of contract, and since in any event electricity is produced by a system that relies on a mix of different forms of generation, governments enforce the contracts through the regulatory system. This is where FiTs, capacity contracts and supplier obligations come in. It is also where political and regulatory risk comes in.

If governments make these sorts of commitments on behalf of customers, the cost of capital falls accordingly, and in turn the economic costs come down, rendering the investments more economic. But even governments raise credibility questions. The German example referred to above is pertinent. In the past, German private companies built nuclear power stations under a system of monopoly provision (no supply competition and no customer switching out of the contracts). As noted above, along came a government (the Red–Green coalition in 2000) which ordered a closure programme without compensation. Then another government came along a bit later and allowed the plants extended lives, but imposed a nuclear fuel rod tax on the basis that the companies would benefit from the revenues of the longer plant lives (which they had been previously prohibited from enjoying, without compensation). A few weeks later the same government decided that they must close some power stations immediately without compensation, not extend the plant lives of the rest, and still pay the fuel rod tax. It is hard to think of a less credible contract, or of a better way of driving up the cost of capital. If a democratic German government can behave in this way, then what protection can investors have elsewhere?

The conclusion is that nuclear can go forward at a reasonable cost of capital only when there is a supportive political framework in place. In the US, the 2005 Energy Policy Act underpinned loans, and gave favourable tax treatment to new nuclear, and the much earlier Price–Anderson Act (1957) provided insurance against nuclear accidents. In Europe the only major country that has met this criterion in recent decades has been France. Yet even here the mask may be slipping. In the 2012 presidential election campaign, the successful socialist candidate François Hollande broke the cross-party political consensus on nuclear and did a 'deal' with the greens to reduce nuclear in France from 80% to 50% of the electricity supply over an unspecified period if he won the election. It remains to be seen whether he now follows this through. But in an important sense it does not matter: investors know that elections come around every few years, and Hollande has broken the cross-party consensus. This spooks investors: the life of a nuclear power station can be up to 60 years, with lots of changes of government in between. It is notable in this regard that in France the main nuclear companies are majority state-owned, and beyond the US many nuclear developments around the world are undertaken mainly by state-owned companies. The political risk is just too great for the private sector.

A small part of an answer, not *the* answer

From a climate change perspective, there appears to be little chance that nuclear will play much of a role in an attempt at rapid decarbonization for at least a decade. Most of the new reactors are likely to be in the Far East, notably in China (it has about 24 under construction). New build in the US is only tentatively beginning to get under way and may not materialize. In Europe there are new reactors being built in Finland and France, and the possibility of new build in Britain and eastern Europe. Japan's programme is, not surprisingly, on hold. The global numbers are small.

These relatively small numbers of new reactors need to be put into context. Many existing reactors built in the 1960s and 1970s are coming to the end of their lives. They will be coming off the systems, and the new build in Europe and the US is unlikely to keep pace with the closures. In Britain, where around 20% of electricity is generated by nuclear power stations, all bar one (Sizewell B) of its 17 reactors are due to close by 2023, and many are already being decommissioned or will begin this process shortly. In Germany, more than 25% of electricity was generated from nuclear at the beginning of 2011. With eight stations closed immediately, and the other nine to close by 2022, there is a very large hole, which new build elsewhere in Europe will not fill. (It will be filled by quite a lot of coal and gas.) Phase-outs in Belgium (seven reactors) and Switzerland (five reactors) will also take their toll on Europe's nuclear output. Uncertainty remains about the future of Japan's 50-plus reactors. Possible life extensions in several cases make these numbers uncertain, but it is not hard to see a quarter of the world's existing reactors coming off line by the middle of the next decade. That is an enormous gap for new build to fill.

The conclusion that emerges is that, whilst nuclear will still play a part, it is unlikely to make much difference to climate change for not only the next decade, but the one after that. Indeed, over this time period its contribution will probably fall in the US, Europe and Japan. After 2030, its prospects may well be very different, particularly if new reactors come along and the cycle can be partially or even completely closed. But this puts most nuclear in the 'future technologies' box that we consider in Chapter 11. In the meantime we need something to fix the problem much faster.

CHAPTER 7

Are we running out of fossil fuels?

All three of the 'silver bullets' – current renewables, energy efficiency and nuclear – rely on the assumption that the conventional wisdom about fossil fuels is right: that we are running out of them and that therefore it is inevitable that the prices will go ever upwards, punctuated by periodic crises. Sadly this assumption is wrong: we are not about to run out, and there is nothing inevitable about rising prices.

As we have seen in the last three chapters, current climate and energy policies assume that renewables will be a good bet, because they will be increasingly cost-competitive against the assumed rising fossil fuel prices. They assume too that since the returns on energy efficiency depend on prices, higher prices mean that more projects will show positive returns. Finally, they assume that nuclear is exposed to direct competition from gas- and coal-fired power stations, and hence again rising fossil fuel prices will make nuclear economic. It is the fossil fuels price forecasts that hold the current policies on climate change together.

Forecasts of high and volatile gas and oil prices are based loosely on the peak oil hypothesis: the claim that oil reserves are known and finite, and that the peak of production has already been achieved, and hence supply cannot keep up with demand. Peak oil is assumed to apply to gas and coal too, so that, the argument goes, we are running out of these fuels fast and, as a result, prices can only increase. These peak assumptions are also applied to a host of other commodities (some of which are ironically used in solar, wind and nuclear technologies) as part of a generalized depletion case, limiting growth potential.

From a climate change perspective, much of this turns out to be at best wishful thinking and at worst nonsense: there are lots of reserves of all three fossil fuels; these are becoming increasingly exploitable; and there are lots of new unconventional sources. The problem is that there is too much in the way of fossil fuels, not too little, with profound implications for climate change policy and the costs surrounding these 'silver bullets'.

The peak oil hypothesis

Peak oil theory is typically associated with the work of geologist M. King Hubbert, who, in the 1950s, looked at the depletion patterns of US oil wells, and made a famous prediction about the date at which US oil production would peak (1970) and then decline.[1] His analysis was based on a well-known set of conventional oil wells, and his followers have extrapolated these depletion patterns in the US to the global oil market.[2]

Although Hubbert got the US peak roughly right for conventional US oil production, as a generalized hypothesis about physical reserves, almost everything that could be wrong with this sort of analysis has, in fact, proved to be wrong. The peak oil hypothesis is little more than a very crude argument *by assumption*. The starting point is that reserves are *known* – that there will be no more significant discoveries. Next it is assumed that recovery rates of the oil wells are known – typically less than 50% – and that technological progress will not alter them much. Finally, it is assumed that demand for oil will keep going up. Armed with these assumptions, the prediction follows. It is an input–output framework, with fixed technology.

Unpacking each of the components, why would anyone assume that the reserves are known? For the century and a half of oil industry history, there have been repeated – and misguided – predictions that resources were about to run out, only to be negated by new discoveries. From the early days of kerosene in Pennsylvania, and the Baku oil wells in the Caspian at the end of the nineteenth century, oil has been thought of as a temporary resource.[3]

This predilection for predictions of 'the end of oil' and imminent shortages has an even deeper past: as noted in Chapter 2, in the nineteenth century, William Stanley Jevons predicted peak coal and the problems of insufficient supply, and the argument has been applied across a range of minerals. The Club of Rome in the 1970s suggested that we would run short of a series of minerals by 2000, and that as the supplies ran out, up would go the prices.[4] Both Jevons and the Club of Rome drew conclusions about the implied limitations for future economic growth, on the assumption of limited substitution. Capitalizing on this gloom, in a famous wager in 1980, the academic Julian Simon bet the environmentalist Paul Ehrlich and his colleagues $1000 that five metals would be cheaper in 1990. He won on all counts. All were cheaper, on average, by 40%.[5]

For the purposes of considering the implications for climate change policy, let's start with oil since it is easier to suggest a peak than for gas or coal. There are three separate but related reasons why estimates of future reserves tend to underestimate the potential: much of the earth's crust remains unexplored; there are lots of unconventional resources; and more can be extracted from existing wells.

Much of the earth's crust remains unexplored

Peak oil theorists started in the US with an underlying assumption that the US's geology had been thoroughly surveyed and was well understood. This was not unreasonable in the 1950s when Hubbert made his famous prediction. It was an assumption that could also have been made about the geology of western Europe.

The remarkable fact is that, in general terms, even this assumption was wrong. Whilst it is true that most of the US's major *conventional* onshore oil fields had indeed been discovered by the 1950s, its *offshore* and *unconventional* reserves remained poorly understood. Indeed, far from oil production peaking in 1970 and then tailing off as Hubbert predicted, over 40 years later it is again on the rise. The tail has been much longer, and in 2011, for example, the US added more extra output than the entire amount lost through Libya's temporary exit from the market during the civil unrest that ousted Muammar Gaddafi. Once the full fossil fuel picture is brought into play (which is what matters for climate change), it is an upward curve stretching for some time to come.

In Europe, the home of geology, the discovery of the North Sea oil and gas fields began with the West Sole find in 1965, and in the 1970s more finds were added. As with Norway, new finds keep being added.[6]

Beyond the US and Europe, the pace of finds slowed down in the 1980s and 1990s for a very good reason: contrary to the predictions at the end of the 1970s, the price of oil collapsed. Yet new finds of conventional oil and gas keep on being added. A quick tour of the globe reveals new reserves off the coast of Brazil, in Canada, off the coast of Israel, in Alaska, in northern Norway, off the Falkland Islands, off the Yamal Peninsula in Siberia, in the Black Sea, off the east coast of Africa and in a number of African countries, and so on. Then there are the large deposits being revealed in the Arctic as the ice melts.

There are also smaller fields that would once have been disregarded. The classic example is in the North Sea where output peaked in around 2000, but where the tail of production has proved more substantial and longer than anticipated. As technology has advanced, and the price of oil recovered, smaller pockets of oil are being exploited, and in aggregate substantive new resources are being added.[7]

These finds are typically offshore and more difficult to exploit. In themselves they would probably not be adequate to meet the rising demand, *assuming* that there are no substitutes for oil. The new discoveries buy time, but if this were all there was to draw on, prices would be expected to rise as the supply–demand balance tightens. Unfortunately for the climate, this is not the main source of additional supplies.

There are lots of unconventional resources

What changed the game was the 'unconventionals'. These are fossil fuels that are found distributed through rock structures, rather than in concentrated reservoirs. The term 'unconventional' is rather misleading: fossil fuels are deposited in a diffuse form as new sedimentary rock is laid down in deltas, estuaries and on sea beds. These sedimentary rocks then get buried under more and more deposits. Over time the oil and gases in these rocks migrate upwards and sideways. Where they find holes in the rock structures, concentrations build up. These are what we refer to as conventional reserves. Thus, below any conventional reserves are usually lots of unconventionals – and typically on a larger scale. So plentiful are these unconventionals that energy expert Daniel Yergin has suggested that 'by [2030] most of these unconventional oils will have a new name. They will all be called conventional'.[8]

The earth's crust has lots of shale rocks (they are the most common type of sedimentary rock), and hence lots of shale oil and gas. The problem has been how to unlock these deposits, and until recently this has been extremely difficult. To release the deposits, the rock structures have to be split open or fractured, which has led to the development of 'fracking' underground, often at very great depths. What makes this challenge all the greater is that it requires the ability to drill horizontally at these depths. Finally there is the problem of finding the most promising pockets of oil and gas locked in the shales, so that the drilling can be directed to the right spots.

To extract shale gas and oil economically therefore requires not just one, but three separate new technologies, which have had to be integrated. Fracking requires water, sand and chemicals to be blasted into the rocks. Reaching the deposits requires horizontal drilling techniques. Directing the drill to follow the best deposits requires seismic information technologies.

Two more ingredients have made the exploitation of these resources economic: access to land and access to water. Land rights are complex and vary considerably between countries. In the US, if you own the surface you own everything underneath it that you can reach from the surface. In contrast, in much of Europe, the state owns the sub-surface. This makes a great deal of difference: in the US, the private incentives are to grab everything below that can be reached from the surface. The early years of the oil industry in the US reflected this: drillers grabbed oil underneath a property, and as much from neighbouring plots as possible – before those neighbours did it to them.[9]

Large volumes of water are used in fracking: it is the lubricant that carries the sand to break open the rocks. It is pumped down the drill hole, and then comes back with the gas. Much of it can be recycled.

There are three major environmental issues – in addition to the water itself – which those whose interests are damaged if gas is abundant and cheap (like renewables and nuclear) have played up in the media as much as they can. These are the chemicals added to improve the fracking process; the protection of aquifers, especially where these are used for drinking water supplies; and methane leakage. These issues are explored further in Chapter 10.

Taken together, the problems of land access, water access, chemical regulation and methane leakage pose significant regulatory issues. But these are not different-in-kind from conventional supplies, and they pale into insignificance compared with the conventional problems of coal. In some cases, shales should not be exploited. In every case, chemicals should be carefully regulated and methane captured. But these considerations are not show-stoppers, and do not justify the outright bans that many green activists demand.

Delivering this combination of technologies on a commercially viable scale became possible by the mid-2000s, and it spawned the shale gas revolution in the US. The scale and speed of development is what makes it a revolution. From a standing start in the mid-2000s, US unconventional gas made up around a quarter of total production by 2011, and it is predicted to rise to a half by 2035. Although the resource base is inevitably uncertain, the US has comfortably enough shale gas to last through at least this century. Even more startling was the fact that some shale gas turned out to be as cheap or cheaper than conventional gas supplies. The result was a collapse in gas prices in the US.[10]

This is enough to turn the US from a major gas importer into a potential gas exporter, thereby triggering a major shock to the whole global gas market, which even the sudden increase in gas demand from Japan after its nuclear accident, and from Germany after its sudden nuclear exit, could only temporarily mask. From a premium fuel, which, as discussed previously, was illegal to burn in power stations in Europe as recently as 1990, gas has now become plentiful, much more secure and, for many countries, relatively cheap. It is easy to forget just how recent the emergence of gas onto the energy scene has been – the same period as that from the beginning of international efforts to address climate change.

Furthermore, it has been found that shale gas deposits are widely distributed – not only in the US, but in China, the Middle East, North Africa (especially Algeria), Argentina, Australia and potentially in Europe too. This has opened up the possibility that at very low cost, gas could replace coal in generating electricity, and given that emissions from gas-fired power stations are about half those of coal, a significant dent could be made in the rising trend in global emissions.

The shale gas revolution is indeed already a game-changer, but it is only part of the story. Shale gas is one amongst several unconventional fossil fuel supplies, including shale oil – again on a potentially vast scale – tar sands, coal-bed methane and tight gas. Given that there is already a lot of coal, now there is abundant gas and oil too. An age of potential energy abundance has arrived, just when the conventional wisdom was turning towards the concepts of scarcity, high prices and peaking production. The unconventionals have turned the reserves position on its head. It is a revolution that is still in its infancy. Coal-bed methane is also abundant, and technologies have been developed to extract it. In Australia projects are being taken forward to liquefy coal-bed methane and export it as LNG to China on commercial terms. Indeed, Australia's coal-related unconventional gas reserves are so great that it could even surpass Qatar as an LNG gas exporter by 2020. Tight gas – gas in sandstones – adds another source. None of these is yet making a significant contribution to fossil fuel supply, but there are good reasons to assume that they probably will.

More can be extracted from existing wells

A second blow to the peak oil theorists is the fact that technological progress not only makes it possible to find and exploit new reserves, and to bring abundant shale gas and oil to market, but it also makes it possible to get more oil out of existing wells. As the oil is extracted, pressure falls and there comes a point where it is more economical to start with new wells than face the costs of lower pressures in existing wells. In most cases, the depleted wells are abandoned.

But as with the shale gas and oil revolution, it is a mistake to neglect technical progress in the fossil fuel industry, and to assume that this is largely associated with the renewables and other 'good' technologies. On the contrary, the pace of oil and gas technology – and its development – is rapid, especially at higher oil prices.

To see the impact of a small increase in recovery rates, assume that the average recovery rate is limited to 50% of its total physical resources. There would then be more oil left behind in existing wells than the entire world oil production to date. Much of this is of course unrecoverable, and many wells have been seriously compromised by the use of water and other means to keep pressure levels up. But the oil remains, and a 1% increase in recovery rates makes an enormous difference to the reserves position. The relevant questions are what technologies are available – or are likely to become available – to increase the potential recovery rates, and at what price they become economically viable to apply.

The demand side

The extrapolation of growth trends in recent years is an obvious baseline for oil demand forecasts. Demand for oil is *falling* in the US and Europe, and so in practice this means taking the growth rate for China and other developing countries and extrapolating it. But is this justified? Is China's phenomenal recent growth sustainable?

The first point to bear in mind is just how recent that economic growth has been. As outlined in Chapter 3, it is only since the end of the 1980s that China has been a growth story, taking up where Japan left off when it collapsed in 1989. It is very much a catch-up story. In energy terms, China was a net oil *exporter* until 1993, and then for two decades it drove up demand for oil and other commodities. Its economic growth was export-oriented, and it exploited both the demand created by the debt-fuelled bubbles in developed countries, and its cheap labour as its rural population migrated to the cities.

For their predictions to materialize, peak oil theorists need this process to keep going – for demand to keep going up. From a climate change perspective, a Chinese collapse would bring emissions relief. Throughout this book we have, however, assumed that it will keep going – that the Chinese economy will double by 2020. But we always need to bear in mind that it might not.[11]

Political constraints

Much of the discussion about oil and gas in the media and elsewhere focuses on the political power of the oil producers' cartel, OPEC, and regional gas suppliers such as Russia. It is argued that what matters is not whether there is enough oil and gas, but where it is. The giant fields of the Middle East are predicted to increase their dominance over global supplies, and to provide the marginal supplies. It is argued that they can hold the world to ransom, by restricting supplies and hence driving up the price. There is then a 'political peak' even if there is no physical peak.[12]

There can be little doubt that in the short run the Middle East has the power to shift prices, particularly Saudi Arabia with its ability to absorb shocks and rapidly increase production. The marginal cost of production is very low, and there is a capacity margin. For as long as this combination of circumstances prevails, both the price level and the volatility will remain problematic and detached from market fundamentals. From the perspective of low-carbon electricity generation and biofuels, the political instability of the Arab states and other OPEC members provides an under-pinning to their investment cases. The political situations in Iran, Iraq, Yemen, Somalia, Syria, Libya, Venezuela and Nigeria are all fragile, and in most cases are likely to remain so. Even stability in Saudi Arabia cannot be taken for granted.

Yet this is not a one-sided problem. Security of supply for oil-importing countries is matched by security-of-demand considerations for oil exporters. The problem for the Arab and other OPEC exporters is that if they restrict supplies too much, they may encourage importers to reduce their demand and look elsewhere. That indeed is what happened in the 1970s and early 1980s. The oil price hikes rendered the Alaskan and North Sea reserves economic, despite their much higher production costs compared with those of the Middle East. The price had its effect – it encouraged other suppliers, and it encouraged energy efficiency measures. As a result, in the early 1980s OPEC faced a crisis on the demand side. Its member states had, however, increased their spending to appease their populations. The 'oil curse'[13] was ramped up by the required budgets, and as prices fell, revenues could be maintained only by increasing output. The incentives to break the OPEC quotas became overwhelming and the cartel cracked open. It would take two decades to get production quotas to stick again. OPEC might not have killed the golden goose, but its actions in the 1970s certainly damaged its dependability.

A similar sort of process may be going on now. The gradual rise in oil prices since 2000 reflects in part the consumer- and debt-driven boom in developed countries. China's export-driven growth has pushed up the demand for oil. When demand is rising, OPEC finds it much easier to cooperate, and the response to rising demand was for it to tighten its grasp on supply. The result has been higher prices. Higher prices have in turn encouraged three responses: the search for new non-OPEC supplies; energy efficiency; and the promotion of low-carbon technologies.

Above we considered the shale gas (and oil) revolution. What made this possible were the technology breakthroughs. But what encouraged the investment were the rising oil and gas prices. The exploration and production investments in the deep waters off Brazil, the US and Greenland are also rendered economic by the high OPEC prices. Developments in Africa are similarly motivated, as are the push for higher depletion rates and the development of smaller fields in the North Sea.

We have seen that energy efficiency also responds to price. Higher prices push up the returns of energy efficiency investments, and make marginal projects worthwhile. The energy ratio can be expected to respond again – indeed, that is precisely what is happening in developed countries now.

Finally, analogous to the 1980s response when nuclear was promoted, the economic returns of low-carbon technologies will have been improved, and the building of wind farms in particular can be seen in this context too. The economics of nuclear have also improved, although in none of these cases is it enough to render any of the low-carbon technologies competitive.

All three responses will permanently damage OPEC, and all three contain the seeds of a crisis for OPEC countries. As oil prices rose in the last decade, they all increased their spending. They needed to: population growth has been phenomenal in the Middle East. The average age in Saudi Arabia has fallen to below 20 years. These young, and under- or unemployed populations pose serious political risks, and indeed the 'Arab Spring' of 2011 was in considerable measure a revolt of the young. More public spending to buy them off is one result – notably in Saudi Arabia.

All three responses involve investments that will play out in the coming decade. In the meantime, the dependency built up in the low-price era of the 1980s and 1990s gives OPEC leverage. Libya's temporary production interruptions in 2011, and the Iranian embargoes of 2012, each sent prices higher. What capped prices in the short run was the global economic crisis of the past few years. This was not caused by oil prices, but it was related. The loose monetary policy – especially in the US – which kept interest rates close to zero and fuelled the consumer boom, came to a shuddering halt. Oil demand in developed countries stopped growing. China still had momentum, but even here the demand for its exports has been affected, with a trade deficit reappearing in particular months, and this will play out in the oil demand figures.

Whilst the combination of the immediate economic crisis in the developed world and the longer-term structural changes in both the demand and supply sides work their way through, OPEC remains hooked on its need for revenue to keep its members from social and political unrest. The OPEC discipline has held up well thus far in the current economic crisis, just as it did in the 1970s, with the convulsions caused by the Iranian Revolution.

Yet the remorseless logic of the market marches on. The shale gas and oil revolutions and the search for alternative supplies will have significant effects on the Middle East. China has turned to Africa and elsewhere for supplies, and as the US moves closer to energy independence, it is conceivable that, in a decade or so, its dependency on Middle East supplies will be reduced more to the margin. Further into the future, Canada and Russia may supply to the Asian markets some of what they had expected to sell to the US. Neither is a member of OPEC.

OPEC's room for manoeuvre is smaller than it may seem. It now needs even more money, faced with its young and volatile populations. If and when the price cracks, OPEC will face enormous pressure to open the taps again to maintain revenues, as in the early 1980s. This creates instability on the downside, and the resulting price effects may be a 'political glut'. When the Saudi oil minister recently announced that the country needed $100+ per barrel to balance its budget, he revealed a dangerous paradox underlying the ratcheting-up of the revenue requirement: he needs higher prices, but higher prices risk setting off a self-defeating process. There is a difference between 'needing' and 'getting'.

It is probably more accurate to regard the idea of political peak oil more as a short-term phenomenon, caused by the time lags in matching demand with supply, creating temporary market power. It is this temporary characteristic that matters in the context of climate change and the reliance on renewables and energy efficiency policies. Over the period to 2020, and indeed beyond, when the oil price is supposed to rise strongly and thereby render these investments economic, the temporary factors may well dissipate. Indeed, the combination of the unconventionals and the new technologies may point towards a very different outcome – with all the political and social upheaval this may cause the Middle Eastern countries. But by then the US in particular may no longer have a vital national interest in the region, as its march towards energy independence goes on, and its dependency on the Middle East weakens. The oil and gas map of the world is being redrafted right under the feet of policy-makers but, at least in Europe, most are stuck in the mindset of the past.

The super-abundance of coal

The physical abundance of gas and oil, and the new technologies that make hitherto uneconomic or unattainable reserves accessible, have undermined the first assumption of the peak oil theorists. There are no immediate physical limits. But even if oil and gas were scarce, the same could never be said about coal. Coal has a dangerous combination of characteristics: it is super-abundant, environmentally dreadful and, if oil and gas prices are high enough, can even be converted to oil.

Because coal is so widespread, and because it comes in different qualities and at different depths and locations in relation to its markets, the full extent of the reserves is unknown. New reserves are being exploited in Mongolia, Indonesia, and Madagascar to add to the abundant coal in Russia, Australia, China, India and the US. Coal has even been exported from Svalbard in the Arctic. In terms of reserves relative to demand, there is enough coal to get well into the next century. The embedded carbon in that coal is sufficient to bring about catastrophic climate change.

Most of the coal that is mined goes into the generation of electricity, and it is important to remember how coal-dependent the electricity systems of the major countries are (see the table in Chapter 2 showing the percentage of electricity generated by coal in a range of countries).

Given the revolution in gas through shale, together with oil and coal, it is very hard to argue that there is any immediate physical constraint on fossil fuels, and in particular on gas and coal. A case can be made in respect of oil, although even here new discoveries and unconventionals reduce the supply risks. Major oil reserves are currently located in a rather small number of countries, many of which have political difficulties (largely caused by the presence of oil and its resource curse). But it has gone relatively unnoticed that even this constraint is, at least in the medium term, limited by the increasing fungibility between oil and the other fossil fuels, especially gas.

Oil and gas are substitutes

The enormous increase in potential gas supply is going to have a greater impact than simply displacing coal in electricity generation. It can be a substitute for oil in electricity generation, particularly in the Middle East. Gas has made inroads into the industrial market, displacing both coal and oil, and also in space heating. It will also create scope for gradually substituting away from oil to electricity.

The exception thus far has been oil for transport – its prime use in the US and a number of developed countries (although heating and, in some countries, power generation are major uses of oil). Road transport, shipping and aviation have remained stubbornly loyal to their oil-based fuels, and their demand has been growing. As China contemplates becoming a car-owning economy, as discussed in Chapter 2, a simple piece of maths translates all those new cars into the demand for oil. Even gains in engine efficiency cannot offset what is a major demand-side effect, although they have been substantial.[14] So far, the only major response has been from biofuels, and as we saw in Chapter 4, this has serious drawbacks.

It is this translation of car ownership into oil demand that gets the peak oil theorists excited. They extrapolate this demand, translate it into a total world demand of in excess of 100 million barrels/day, and then set this against their assumed peaked supplies, and get a massive energy crunch.[15]

But what they never do is imagine a world with all these cars and a *falling* demand for oil, because they do not think through the electrification of transport or the direct fuelling of these cars by gas. All they see is the simple equation of more cars equals more oil demand. If electric cars become commercially viable, or gas is substituted directly for oil, then gas can take up some of the transport demand. Hence gas may be a substitute for oil.

Such a transition will take time. Electric cars may already be on the roads, and there may be the beginnings of rudimentary battery-charging systems in some European cities, but there will inevitably be many more non-electric cars between now and an electric future. Yet even here, there are transition technologies – hybrid cars, and indeed gas-fuelled cars and trucks.

Finally there is gas-to-liquids (as well as coal-to-liquids): oil can be made from gas (and coal), and at the now extremely low gas prices in the US, this is becoming economic. Were gas prices to remain so low (which they may not of course), and were the sorts of oil prices that peak oil advocates predict to come about, this technology would become widespread, so that gas would not only be a substitute for oil via electric cars, but also be the fuel source for conventional cars.

The price of fossil fuels

The peak oil theorists' claim that production has peaked whilst demand continues to rise strongly is highly contentious. The discussion above suggests that the world may turn out very differently – or it may not. These future prices are inherently uncertain, and it is this uncertainty that shapes the future policy context. Unfortunately, the peak oil theorists (and many politicians) *assume* that they know what future oil and gas prices will be. Indeed, they need to know in order to be able to pick the winning technologies.

The argument goes that because oil and gas prices are going to rise, and because they are also going to be volatile, the task for policy-makers is to protect consumers from these high and volatile prices by making sure that technologies are invested in which do not depend on fossil fuel prices – notably nuclear and renewables. Ed Davey, as British Secretary of State for Energy and Climate Change, for example, believes that the aim should be to 'insulate' the economy from fluctuations in world gas prices: 'every step the UK takes towards building a low-carbon economy reduces our dependency on fossil fuels, and on volatile global energy prices', ensuring that 'our economy does not become hostage to far-flung events and to the volatility of market forces'.[16]

We have been here before. The same certainty about prices was trotted out at the end of the 1970s when oil prices peaked at $39/barrel (around $140–$150 at current prices). This time around, there are remarkably similar types of Armageddon predictions, and forecast price increases are used to justify policies that governments and lobbyists may wish in any case to pursue.

This certainty blinds governments to other possibilities. Suppose prices fall. This would give considerable economic relief in the context of the global economic difficulties, but it would also be a serious political challenge, and it would undermine the economics of the technology winners that politicians have picked. Instead of wind, solar, and indeed nuclear becoming economically competitive against fossil fuels, they would remain in need of *permanent* subsidy. The current narrative is that support for renewables is a transitionary affair, for immature technologies that after around 2020 will become cost-competitive with all the other technologies, supported only by a carbon price. As the Department of Energy and Climate Change puts it, 'our long term vision is a market where low carbon generators compete fairly under a robust and stable carbon price'.[17] A similar view appears now to be emerging from the European Commission.[18] But in energy policy the temporary has a nasty habit of becoming permanent, and that is what may well happen. This in turn would probably provoke a political reaction from voters who had been told that renewables would be cheaper, and this then might reduce the incentives for governments to maintain the subsidies that consumers have been forced to pay. Indeed, the political commitment to subsidies has already begun to crack.

The industrial reaction might be worse. Countries such as the US, which have embraced unconventionals like shale gas, already have a very considerable advantage over economies with relatively expensive renewables. In the US, the gas price has been as low as $2 per million British thermal units (MMBtu). Europe's prices are more than four times higher. The competitive effect here goes indirectly through industrial costs, and it could be potentially very adverse for the Europeans and even the Chinese. The US gas price is now so low that in some circumstances it can outweigh China's (diminishing) cheap labour advantages. The result is an emerging repatriation of energy-intensive industries back to the US, notably in petro-chemicals now that it has this cheap and abundant feedstock at home. For Europe, hampered by more expensive labour costs, the gas price differential with the US may be a heavy blow to its competitiveness. This will have the incidental effect of lowering European carbon production still further, but of course this makes no difference at all to global carbon emissions. The only gain in climate change terms might be that European economic growth is even lower, thereby bearing down on incomes and hence carbon production and consumption. Yet even this will probably be offset – by higher growth in the US.

A world of abundance

A world of abundant fossil fuels is not good news for climate change. Cheaper energy sources would continue to provide a competitive advantage to countries that choose to free-ride on others (see Chapter 8 for a discussion of the free-rider problem). It would force politicians to explain that the transition from a carbon to a low-carbon economy will be expensive, and it will require permanent subsidies, or a much higher carbon tax. It makes a nonsense of the claims about a low-cost transition to a low-carbon economy based on current renewables, energy efficiency and even nuclear.

Yet in all this energy abundance, there is one silver lining: fossil fuels emit different quantities of carbon when burnt. We have noted that emissions from gas are around half those of coal. Thus, whilst the overall contribution of fossil fuels may remain high, a switch from coal to gas, especially in electricity generation, would make a big difference to aggregate emissions. Gas may therefore provide a *transition* fuel and limit the growth of coal. And it might therefore buy time. This turns out to be a major opportunity, which we will come back to in Part Three. But before we consider this, the final part of the current policy framework needs to be assessed – the Kyoto Protocol and the largely futile efforts to get a global agreement on carbon production caps.

CHAPTER 8

A credible international agreement?

If specific technologies, and the associated policies, are not sufficient to bridge the gap between aspirational targets and the carbon reality, what about the overarching framework within which they are set? Is Kyoto an appropriate framework? Has it contributed anything substantive to the reduction of emissions? Is it ever likely to? And if not, why not?

It is now two decades since the world came together at the Rio Earth Summit, and signed up to the 1992 UNFCCC. The science of climate change was then quite new. The IPCC had been set up two years earlier. These were heady days: the new global threat would be met by nations uniting for the common global good. The ultimate objective of the Convention is to stabilize greenhouse gas concentrations 'at a level that would prevent dangerous anthropogenic (human-induced) interference with the climate system.'[1]

But agreeing what exactly should be done, by whom, and by when, turned out to be altogether more difficult – indeed, in many respects, practically impossible. At Rio+20 in 2012, it was painfully obvious that the hopes had not been translated into happy outcomes. The UNFCCC treaty set no mandatory limits on emissions for countries and has no credible enforcement mechanisms. Rather, it set about creating an inventory of emissions for the 1990 baseline, and set up a process of meetings of the parties from 1995, and the path to what seems like endless summits, where NGOs gather with government ministers and make profound statements, but typically achieve little relative to the scale of the challenge.

To the extent that any significant progress was made at all, it was around the core idea that countries should adopt binding carbon-production targets. Thus was born the Kyoto Protocol: adopted in 1997 in Kyoto, entering into force in 2005, with the first commitment period set for 2008–12, a decade after it was first agreed.

Once the debate turned to concrete national targets, the self-interest came to the fore. Most countries opted out of any targets at all, preferring to leave it to others – under the politically convenient principle of 'common but differentiated responsibilities'. Carbon targets are a classic case of free-rider incentives – let others do the heavy lifting, whilst sitting back and enjoying the fruits of their labour. When it comes to climate change, self-interest and collective interest are very different. To understand why Kyoto failed to deliver what it promised, it is necessary to first understand these incentives.

The prisoner's dilemma: why agreement is so difficult

It might seem somewhat flippant to treat climate change negotiations as a 'game', but game theory does in fact provide a very good way of characterizing the problem and identifying why it is so hard for countries to agree.[2] Specifically, climate change negotiations resemble a particular type of game called the prisoner's dilemma. In this stylized scenario, two isolated prisoners under interrogation face the following problem: the inability to reach the cooperative outcome and thereby get lower sentences for both of them, given that the incentives of the game dictate a sub-optimal outcome if they pursue their own narrow self-interest.

The game is set up as follows. The police don't know who committed the crime and they want a conviction. The prisoners have two options: either confess or do not confess. The pay-offs are as follows. If one confesses and offers evidence against the other, and the other does not confess (by staying silent), the confessor will be freed. The other gets to spend a long time in prison, since the police now have the evidence they need to convict him. If neither confesses, the police cannot make the charges stick and they both get a short sentence. This is the best overall outcome, but it requires them to cooperate with each other; otherwise the risk is that each faces a very bad outcome, since not confessing when the other does results in a long prison sentence. So both go for the safe option: confess and betray the other one, and both get imprisoned for a medium amount of time as a result, but not for as long as if each had stayed silent whilst the other confessed. Staying silent is a risky business. So the only way of achieving the best overall outcome is to cooperate with each other and stay silent, but they can't and anyway they couldn't trust the other to deliver on any agreement. That's the dilemma.

The prisoner's dilemma neatly translates across into a climate change dilemma. If every country adopts binding targets, all would face costs, but the world would be better off than if nothing were done. This cooperative solution is equivalent to staying silent in the prisoner's dilemma. Like the prisoners, the countries are each taking their own decisions independent of the others. But what if some countries adopt binding targets, but others do not? The ones that don't do not face lower costs (they don't have to take any mitigating actions now), but still suffer a bit of global warming. The ones that do adopt the binding targets have to face the costs of acting now and the fact that the impact of their emissions reductions will have less effect because of the emissions of those that won't play ball. For them it's a bad outcome to agree to caps, and let others free-ride at their expense. So the default outcome is that no one takes on the emissions reductions. Consider the US and China, and why neither took on binding emissions caps.

This very simple characterization turns out to be extremely robust against changes in the assumptions. In a nutshell, this is the situation with which international negotiators are faced. Once this game is made more realistic, the incentives to free-ride may get even bigger. The most obvious complication is that the costs and benefits vary considerably between countries. We have seen that some countries may actually gain from some initial increases in temperatures. Russia gains a lot more continental-shelf access to oil and gas in the Arctic, and possibly more agricultural land. Its heating bills might go down. Similar considerations may apply to Canada as the ice retreats and its enormous mineral resources become increasingly accessible. Others are not likely to be so lucky.

How then might such an incentive problem get solved? The obvious answer is to enter into a formal agreement. Our climate prisoners need to formalize a deal. But is not hard to see how difficult this might be. Not only is it against the narrow interests of at least some countries to join, but even if they did they need to be sure that others will not cheat as the prisoners surely will – i.e., agree but then not fulfil their obligations. So the incentive to agree depends on the analysis of what would happen if anyone cheats.

This problem is similar to forming a cartel and awarding production quotas to each member. Think of the example of OPEC trying to fix the oil market, as discussed in the previous chapter. In the OPEC game, the constraint on each producing country is the need for revenue to appease its population. The equivalent constraint in climate change is the political and competitive need to keep energy prices low to appease voters and protect business interests. OPEC and climate change share other dimensions of the game. How do you monitor compliance? How do you detect cheating? How do you enforce the agreement by punishing cheats? OPEC struggles to deal with the monitoring, and it has no enforcement mechanisms to hand. Hence its 'successes' in monopolizing the oil market have been limited.

The problems of the Eurozone provide another interesting analogy. The countries signed up to the Stability Pact, agreeing maximum deficits. Then a shock came along, and even Germany violated the agreement. Greece cheated on the measurement, monitoring was weak, and enforcement non-existent. Climate change is also vulnerable to shocks. A shock (like an economic recession) creates different priorities; emissions by countries are not always easy to measure and verify; and cheating is obviously possible.

In the first Kyoto commitment period, for example, Canada was way above its emissions cap, and whilst there is provision to catch up in the second period (plus an additional penalty reduction requirement in the next period), Canada has since effectively given up. Worse, where there is money flying around to pay for emissions reductions in developing countries, the scope for corruption is considerable and, perhaps not surprisingly, has been exploited.

It is hard to think of a major global public policy issue that provides more incentive problems. And this is before consideration of the crucial difference in this game and that of our prisoners – the damage is in the future, not the present, and hence is inflicted on future generations. In the climate change version of the prisoner's dilemma, the prisoners are not necessarily even in the room, or even the building.

All sorts of 'solutions' to the game have been suggested. Some argue that the catastrophic impacts are more immediate than people think and that everyone will be a loser. In other words, it is claimed that the structure of the pay-offs is such that it is in everyone's interest to act, regardless of what others do. Nicholas Stern comes close to this in his discussion of China's incentives.[3] The argument is that China will be badly affected by climate change because its water supplies from the Himalayas will be radically reduced, and its deserts will grow. Yet even in this case we need to be careful in extrapolating into the future, and very careful about the uncertainties of the science when it comes to predictions not only about the climate in general, but also about the specific effects in specific regions. China too will have benefits – it is currently very cold in parts of its territory in winter. The costs of action might fall very heavily on its immediate economy, and its political leaders may be more concerned about the short-to-medium-term budgeting needs to keep the population under their authoritarian rule. With a fast-growing economy, future Chinese are likely to be much wealthier than the current population, hence the trade-off between imposing costs now on the relatively poor, and costs incurred by richer people later, is much sharper than for slow-growing European economies. Finally, China alone cannot solve climate change; it needs others (also faced with the same dilemma) to act.

Others point to moral and educational considerations, arguing that voters will decide according to broader motivations than national self-interest. We saw how such ethical considerations might include rights and responsibilities, on a country or individual basis. But politically, this is a big ask and we do not have enough time for a programme of moral re-education, even if it were desirable. There is little evidence that voters are putting the future much above the crisis-ridden present. Interest in climate change appears to be a luxury good – fine when economies are doing well (as long as it does not cost too much), but not so pressing otherwise. And, in any event, most governments are not democracies. China is ruled by an authoritarian elite as, effectively, is Russia. If the future of our climate rests on the ethical outlook of Vladimir Putin and the emerging Chinese leadership of Xi Jinping and Li Keqiang, then there can be little grounds for optimism. Worse still for this 'democratic' argument, the US seems even less interested in climate change than some of the dictatorships. This may be true for Canada too. This is the hard political reality within which climate change needs to be addressed, not some ideal-ized world pervaded by 'goodness'.

What these considerations from game theory tell us is that climate change is not an obvious candidate for a globally binding agreement. Indeed, it has virtually none of the features that an effective global carbon cartel has. No wonder, in David Victor's terms, climate change negotia-tions are 'gridlocked'.[4] But this did not stop the UN trying to broker a deal, or the EU promoting the Kyoto framework in the belief that if the Europeans just show enough 'world leadership', it will all work out fine.

Kyoto and its free-riders

The Kyoto Protocol tried to disaggregate the agreed overall target of reducing global emissions by 5.2% below the 1990 baseline by 2012. Developed countries were set individual carbon production caps. In the sunny economic climate of the 1990s, emissions were on a rising trend, and hence the 5.2% represented a much greater reduction against predicted business-as-usual figures. All countries would enter into an agreement to act eventually, but only some would have to act now. The Protocol would set an initial period (2008 to 2012) in the expectation that there would then be a second and third period and so on, negotiated as experience was gained along the way.

For the initial period, countries were divided into two categories: Annex 1 (developed countries) and Non-annex 1 (developing countries). Annex 1 countries would take on binding production caps immediately, whereas the others would take 'measures', but not apply binding caps now.

The Kyoto Protocol was conditional: it would come into force only if enough countries, and emissions, were covered – at least 55 countries in the UNFCCC process, and 55% of global emissions in 1990. Because the US did not ratify (because China did not have a cap), the criteria for coming into force looked like they would not be met, and so there was the possibility that it would not get off the ground. The Europeans – for reasons discussed below – were very much in the lead, and they recognized that if they could get Russia to ratify then the criteria would be met.

Russia played hard to get, showing just how powerful self-interest is in international climate change negotiations. It wanted side payments – notably help in joining the World Trade Organization (WTO) – and a deal which would not actually cost it anything. Since the Soviet Union had collapsed at the end of the 1980s, there was a lot of 'hot air' to play with. Its collapsing energy-intensive industries meant that its emissions would fall significantly anyway. 1990 was therefore a convenient base year: emissions were bound to fall in the 1990s. The trick was to get paid for something it would achieve in any event. This, along with the WTO, formed the 'bribe' that the Europeans offered to get it on board, and Russia duly obliged in 2004.

The US position was not as self-centred as some have sought to portray it. Although Vice President Al Gore had actively participated in the Kyoto negotiations, and President Clinton signed in 1997, the US Senate refused to ratify it. Not a single senator voted in favour. President Bush pulled out entirely in 2001, and President Obama remained resolutely on the outside. Thus from Clinton to Obama, the US position has remained remarkably consistent.

The US position – in effect, that it would accept caps only if China did so as well – had a core rationale: the great growth in emissions from 1990 would be driven by China. As long as China kept expanding its coal-burn, emissions would go on rising. Climate change since 1990 has been largely about China and coal. The US argument was therefore a very rational one – grounded in an all-too-realistic assessment of the causes and paths of emissions.

China, of course, could argue that it was developing – income per head was much lower, and emissions per head were correspondingly lower. This was a powerful moral argument, but it could not avoid the fact that if China pushed on towards US levels of emissions per head, the resulting carbon concentrations in the atmosphere would be much greater – regardless of what the US or others did. The global climate could not afford this development path, irrespective of the moral arguments, and irrespective of what the US did.

What made China's position even worse from a US perspective was the fact that its economic development depended on US demand for its exports. Thus if the US faced a cap and China did not, China's exports would have an unfair advantage over home production in the US. Understandably, the US argued that it was both or neither; China argued that it was the US only that should adopt binding caps. Thus was Kyoto neutered from the outset, and it never recovered.

A plausible case may have been made that the US and Europe should pay China not to develop its coal-intensive industries, but a moment's reflection illustrates how politically unacceptable that would be. It could be portrayed as the US transferring chunks of its wealth to the Communist dictatorship of China in order for China to develop a more competitive economy to take on the US. Combine this with the popular claims in the

US that China had been manipulating its currency, that it had used trade restrictions and displayed a remarkable lack of regard for intellectual property rights, and it is not hard to understand the US position. Add to this the growing military competition in the South China Sea and it is easy to see that paying China to reduce emissions on anything other than the margin was never going to happen, and it further illustrates why a top-down framework like Kyoto was ill-fated from the start.

Experience of mechanisms within the Kyoto approach gave yet more credence to the US position. The Clean Development Mechanism (CDM) provided for money to be transferred to developing countries via the Certified Emissions Reductions (CERs). The idea was that this would enable developed countries to help emissions reductions in developing countries, notably where these were cheaper than emissions reductions in developed countries. What happened? China has been one of the main recipients, in large measure for projects destroying the greenhouse gas hydrofluorocarbon (HFC-23), which it probably would have done anyway. Even more cynically, it taxed the CDMs too. If China could not be trusted to use these limited resources for the *additional* emissions reductions they were intended for, why would an even larger fiscal transfer work? Americans were not fooled.

The result of the US refusal to ratify, and the absence of a Chinese cap, meant that Kyoto defaulted to a de facto European treaty commitment and, in effect, Europe adopted what looked more like unilateral carbon targets. Kyoto was well suited to this purpose. Europe was deindustrializing anyway, and the major structural adjustments following the collapse of the Soviet Union further reduced its energy intensity.

So Europe could take the moral high ground, proclaim its rectitude, and claim that it would lead the world in tackling global warming. Its leaders felt content in this role: they could strut the world stage in the sure knowledge that they would probably meet their targets anyway, even without the global economic downturn and the Eurozone economic crisis. Tony Blair epitomized this leadership, notably at the famous Gleneagles summit, although Merkel the scientist could claim that role too.[5]

Copenhagen collapse

The trouble with the global leadership argument is that Europe risked being found out. What would happen if, after investing so much political capital, it turned out that others were not excited by the arguments and refused to play ball? This is something that the architects of the European position seem never to have thought seriously about in the run-up to the Copenhagen climate conference in 2009.

The hype around the run-up to Copenhagen was extraordinary. Ambitious goals were announced, and the EU offered to raise its 20% reduction by 2020 to 30% if others joined in (and in the process revealed how easy the 20% target was for Europe). The election of President Obama in 2008 had led environmentalists to assume that the US would come on board in the mistaken belief that the US opposition was generated solely by President Bush and the Republican Party. They conveniently forgot that Bill Clinton, a Democrat, had been unable to deliver ratification of the Kyoto Protocol. The green NGOs and the European politicians gathered at Copenhagen could not admit that there was anything flawed about Kyoto: it had to be simply a matter of political will. For some, the world was divided into 'the good guys' (the Europeans) and 'the bad guys' (the US). Bush provided the ideal target for the greens and the Left: Republican, responsible for the invasion of Iraq, and American.

Copenhagen represented almost everything that is wrong with global climate change summitry. The runway at the airport was full of the aircraft of all those who wanted to be in on the conference, from rock stars and fading politicians to the European Commission at its most hubristic. A veritable circus developed, with green NGOs coming to the fore. For many of them it was the place at which to be and be seen – for political gestures and self-importance. NGOs had their 'stands' and did their best to recruit. The media duly obliged.

This circus – and the extraordinary expectations, given the underlying political realities – made negotiating an agreement close to impossible.

The first week of the conference was devoted to grandstanding and media hype, with exaggerated claims about imminent climate catastrophe. Only towards the end did the key world leaders arrive, and by then the chaos had reached fever pitch. Agreeing a text was impossible: the Europeans expected Obama to sign on the dotted line, with US production caps and a commitment to an emissions trading scheme. The Chinese were expected to do their bit too.

What happened was predictable to all but those in the cocoon of the conference. As is the way with these sorts of international conferences, the negotiations go right to the wire, and because everyone knows that this will happen, no one gets serious until the last moment. And it was right at the end that it became abundantly clear that the EU dream was not going to materialize. Nobody wanted a complete collapse – world leaders do not attend funerals until the patient is really dead. So at the last minute something had to be cobbled together from the wreckage.

That something was deeply wounding to the Europeans. The US and China, supported by Brazil, India and South Africa, cut a deal without consulting the conference, and presented it as a fait accompli. President Obama and the Chinese leadership then simply left. All that the conference could do was to 'take note' of what was called the Copenhagen Accord. Countries could subsequently submit emissions reductions pledges or mitigation action pledges, all non-binding. The Europeans were not even at the table. Europe's aspiration to world leadership ended in humiliation.

What was salvaged was anything but ambitious, despite the spin. Summits and conferences almost always end up with 'agreement', and spin doctors go into overdrive to put the best gloss on the outcome. All that really came out of Copenhagen was an ambition to limit global warming to 2°C, and a wish list of national ambitions to cut emissions. The European Commission, in its own assessment of the outcome, concluded that the sum of all the voluntary commitments in the so-called Copenhagen Accord, even if achieved, would not meet the 2°C target.[6]

Reality bites at Durban

If anyone expected the cold reality exposed at Copenhagen to lead to a reassessment of the Kyoto approach, they were to be disappointed. The Europeans kept plugging on as though Copenhagen were merely a temporary setback, and the focus of attention shifted, via a gathering at Cancun in Mexico in 2010, to Durban in South Africa at the end of 2011. The first commitment period for Kyoto was to end in 2012: there was still time to cut a deal, and Durban represented the last chance to push on in 2013 with a more demanding and robust second commitment period.

The conference in Cancun in the intervening period was low-key – low expectations, and little progress. Focus shifted from the big question – legally binding caps for all the main emitters – to other dimensions, notably creating a global climate change fund, and the REDD (Reducing Emissions from Deforestation and Forest Degradation) forestry programme.[7] These were important issues, but not the main deal.

Durban was in one sense a good choice for the next make-or-break conference. South Africa provided an example of the many dimensions that needed to be taken into account in the negotiations. It is heavily coal-dependent, as explained in Chapter 2, and its population is rapidly expanding. Its agriculture is depleting the carbon in the soil, and it wants to grow fast to be on a par with the economic development of the other so-called BRICS: Brazil, Russia, India and China.

These causes of carbon emissions epitomized by South Africa – coal, growth, and population – were not, however, the main focus of the Durban conference. Rather, the Europeans tried yet again to push the US and China towards legally binding caps. This time there was at least some recognition of the problems. The US was not going to move in the shadow of a presidential election due in late-2012, especially with the Republicans having broken through in the 2010 mid-term elections. In neither of the main US political parties was there much appetite for addressing climate change. The Chinese, wary of increasing risks to their economic prospects, were reluctant too, and in particular reluctant to commit to anything that might be legally binding. They did, however, want to avoid the blame for failure. The US could hide behind China's failure to come on board, and China could point the finger at the US. The climate change version of the prisoner's dilemma was there for everyone to see.

Perhaps more ominous was the threat of withdrawal of three key players, none of which now wanted 'more Kyoto'. Japan, still the world's third largest economy, had suffered the nuclear disaster at Fukushima. It would need much more in the way of fossil fuels to compensate for the setback to what had been assumed to be a nuclear future – and hence a low-carbon one. Canada was now one of the world's largest holders of oil and gas reserves, including the tar sands. Climate change mitigation not only threatened its markets, but climate change itself was opening up major new reserves as the Arctic warmed. Russia too benefited from the Arctic warming, and given its vast oil and gas reserves and its over-whelming budgetary dependence on the revenues, was not much inter-ested in mitigation. The upshot was that the US and China (and India) would not agree legally binding caps, and Japan, Canada and Russia were even less willing participants.

The second commitment period after 2012 had therefore effectively lost three key players, and was increasingly just an EU venture. EU politics still dictated a 'leadership' role, but even here doubts were beginning to surface. For the first time, it dawned on the Commission that others might not follow. Then what? The days of easy economic growth were over. Mass unemployment had returned to Europe. Taxes were rising. Ordinary people had begun to lose interest in climate change, and in particular their willingness to pay for it.

Carbon leakage becomes a much sharper problem when it means the announcement of job losses. In Britain, George Osborne, the Conservative Chancellor of the Exchequer, stated in 2011 that he was not going to 'put Britain out of business' if others did not follow.[8] In the Netherlands, a right-of-centre government backtracked on offshore wind, and subsidies have already begun to be cut in Britain, Spain and Germany. But for the possibility of a continuing stranglehold on future potential coalitions by the Green Party in Germany after the 2013 elections, the European position itself might have buckled in Durban. Conveniently perhaps, since there was never going to be a legally binding agreement at Durban, the Europeans could still take the moral high ground, safe in the knowledge that others could then be blamed.

The eventual outcome, after the conference had been extended for a couple of days, was predictably spun as 'a breakthrough'.[9] It was anything but. All that was salvaged was an 'agreement' to keep trying to reach an agreement, with a target not for emissions reductions, but for reaching an agreement. It would be hard to make this up. This lofty ambition is to be achieved by 2015, and even then will apply only from 2020. This was how the Europeans kept Kyoto on a life-support system – just. The second commitment period would happen, and for the Europeans that meant that the flagship EU ETS could struggle on (since the Kyoto production caps provided the quantities against which the permits could be issued). But that was it, and the participants could not even agree on what the proposed agreement's precise legal force would be in 2015, and had to compromise on this too.

Even if caps in 2020 were to be agreed, the cold reality of Durban was that the major countries agreed to do very little *for another decade*. In that period, the Chinese and Indian economies may well double in size, with all the associated emissions, notably from coal. What Durban really symbolized was an acceptance that the 2°C target was no longer feasible. Of course, this did not figure in the ludicrously triumphalist spin that European politicians put on the outcome.

No realistic prospect of an international binding and effective agreement

In the way of summits and international conferences, although those concerned with climate change have now gone away and started the foundation work for another round of negotiations, what will happen is that they will focus on the final day of the prospective summit in around 2015.

The sad fact is that most politicians will now give priority to other pressing problems: the economic crises, bank failures, the Eurozone, and keeping the Chinese economy going. In the US case, this includes promoting energy-self-sufficiency through shale gas and oil, and drilling for conventional oil offshore. At the day-to-day level, in democratic countries, these same politicians will also be keeping one eye on the next election (i.e., holding on to their jobs), and if climate change is not a priority for constituents, it isn't likely to be high on the agenda.

By 2015 the world economy could be mended. Growth might be back on track. Wealth may be going up again, people might find jobs and budget deficits may be back under control. If (and it is a big if) the world is in a good, or at least improving, economic place, the negotiations might start to make more progress.

Even if such a benign outcome were to materialize – and this would currently appear to be rather optimistic – it has a darker side for the climate. The very growth itself will have made the problem a whole lot worse. Emissions will have gone up, and a lot more coal will have been burnt. Furthermore, one reason why the growth might return is if fossil fuel prices decline. The booming US and Canadian oil and gas sectors – and all the jobs they are creating – may be one reason why recovery takes place. That is not necessarily good news for climate change.

Political leaders like to be seen to agree, and no doubt in 2015 some sort of 'agreement' will be forthcoming. But the remorseless logic of the climate change game dictates that it is unlikely to be one that makes more than a marginal difference. The players in the game face different pay-offs, and they cannot rely on effective enforcement. Free-riding is built into the very structure of the game.

The realistic conclusion is that there is not going to be any *serious*, legally binding, international, and enforceable, deal for at least a decade, and possibly never. The parties will go on talking, and the showcasing of international events will go on too. But this is unlikely to solve global warming. For a solution we have to look elsewhere. The practical question is whether these Kyoto-based summits are ultimately even positive. By carrying on a largely fruitless process for years, the need to explore alternatives is not given the priority it deserves. A credible Plan B struggles to get off the ground, whilst world leaders continue to chase the deeply flawed Plan A. Indeed, the negotiating process itself becomes a barrier – and an excuse – for not taking other options seriously. That is the real tragedy of Copenhagen and Durban.

PART THREE

What should be done?

CHAPTER 9

Fixing the carbon price

Let's take stock of the argument so far. The current approach to climate change is not working, and it is not likely to work any time soon. Emissions will keep going up, and it will be another decade before anything substantive happens on the Kyoto front – if at all. In 2020, some 30 years after the baseline for the agreed targets for carbon reductions, the Durban agreement holds out the prospect that there may eventually be some sort of legally binding agreement. It is a hope, but not a guarantee. By that time the carbon concentration may have crossed the 400 ppm threshold on current trends. The IEA's forecast at the end of 2011 indicated that there is practically no prospect of holding global warming to 2°C, even if current policies were fully implemented.

Top-down global caps are not leading us towards a solution to climate change. To carry on along current lines is to condemn future generations to a much hotter world. We will bequeath them an unsustainable world. Instead of being good stewards and maintaining the environmental assets intact, we are steadily undermining their prospects, perhaps even catastrophically.

It is easy to despair. Indeed, arguably, that is what public opinion increasingly reflects. The pessimistic line has considerable plausibility: more coal; fast economic growth in the developing world; and another two billion people together add up to a wall of consumption that looks like remaining largely carbon-driven. The Kyoto architecture is seriously flawed. Since no one else is doing much, why should we? The classic free-rider problems are very much alive and kicking.

Many will conclude that there is only one option – adaptation. For those in the temperate zones, for at least the next two to three decades, it might even be a case of lying back and enjoying it, and hoping the climate change problem goes away as new technologies come to the rescue. Up to 2°C is after all not all bad.

Some have even come to doubt the science, in part driven by scepticism of the green NGOs' alarmism. There is a history here: in the 1970s, a new ice age was forecast and the Club of Rome focused on the finiteness of resources and the implications of their depletion. The peak oil advocates, backed up by environmentalists, tell us that the end of oil is already upon us and add to the alarmism.

An alternative, more positive, approach is to turn the problem on its head: to start bottom-up with national and regional approaches, and to use three key policy instruments – the carbon price, the gas transition, and R&D – rather than place overwhelming emphasis on regulation, wind and solar renewables, energy efficiency and Kyoto. Current renewable technologies have their role to play, but they are not sufficient, and the sad fact is that there are limited resources, which need to be deployed as cost-effectively as possible. There simply is not enough land for the wind turbines, or water for the current biocrops, to close the gap, even if they were cost-effective.

In this third part of the book, these three components of a more cost-effective – and optimistic – strategy are laid out. The first, which this chapter considers, is carbon pricing.

The economic approach

Economists have long argued that prices play a very important part in the allocation of resources. They also focus on the fact that resources are limited: to spend on one thing is to decide not to spend on another. Although patently obvious, it is repeatedly ignored by politicians, who try to avoid such painful trade-offs. Prices transmit information about costs, and hence provide the mechanism by which markets allocate resources between competing goals. The corollary is that, without prices, resources are likely to be misallocated, often (as in the case of carbon) massively so.

For economists, the rationale behind a carbon price is consequently also obvious, but it needs to be restated for the very good reason that many environmentalists prefer direct intervention and command-and-control. As has been noted, many in the green camp are also on the Left, and some combine a socialistic preference for planning with their green objectives. Indeed, for some there is an authoritarian streak: they 'know' the answers, and it is just a question of forcing people to follow their prescription. Those who beg to differ become 'the enemy' and hence have to be discredited. John Sauven, Executive Director of Greenpeace UK, reflected this intolerance when he was quoted at the end of the Copenhagen climate conference: 'The city of Copenhagen is a crime scene tonight, with the guilty men and women fleeing to the airport'.[1]

The economic case starts from the opposite pole: it is based on uncertainty, not certainty. It assumes that we do not know the costs or people's preferences – and hence the conditions of supply and demand. The contrast between those who 'know' which technologies are the right ones and those who promote the economic approach could not be greater. If information is perfect, there is no point in bothering with the market. 'Winners' can be picked and imposed on the economy.

Economists do themselves no favours when, notwithstanding this case for carbon pricing based on uncertainty, they nevertheless build large-scale integrated assessment models (IAMs), which attempt to roll the science and the economics together, and then make fairly precise claims about the costs and benefits of alternative policy paths. This approach is built on 'general equilibrium models' of the economy, which are ultimately based on *assumed* exogenous preferences of *assumed* rational utility-maximizing individuals, and *assumed* exogenous technology. These are the assumptions upon which the predictions are made, and one of the outputs is the calibration of what the price of carbon should be to achieve a given mitigation path.

Such models are questionable in any economic context, but they are doubly so when it comes to climate change. For they take as given a great deal upon which to base forecasts and projections decades into the future. It is what lies behind the Stern Review's statements about the costs and benefits of early action, and the IPCC's scenarios and predictions.

One might think that past failure might induce some humility on the part of these economic modellers, but quite the contrary. Looking back over energy forecasts is a sobering exercise, and it is notable that the premier energy modelling and forecasting body, the IEA, does not do backcasts – i.e., exploring its past performance. If it did, there might be fewer confident PowerPoint presentations of its 'latest' scenarios.[2]

The market is, of course, far from perfect. There are many market failures (and many governmental ones too). So relying on markets does not mean zero intervention. What matters is giving the right signals and being prepared to react and revise as new information becomes available. Of the many market failures, the absence of a price on carbon means that it will be left out of the calculations of companies and individuals. If you do not have to pay for the consequences of your carbon footprint, it is unlikely that you will do much to try to limit the damage you are causing. The emissions associated with your consumption will be external to the market – an externality. In order to rectify this failure, a carbon price needs to be set so that it is internalized – and then it has to be taken into account. There is no option but to do so.

We do not know how best to cut emissions. Our knowledge is necessarily incomplete. So we can either guess and then pick winners from the available technologies, or we can set the price of carbon and see what happens. Suppose we do the latter. Putting aside for a moment the question of the level at which the price should be set, once it is in place it changes the costs that each household and company faces. Now carbon-intensive consumer goods are more expensive compared with the low-carbon alternatives. Coal will be more expensive than gas as a fuel for electricity generation. So both consumers and producers will adjust their behaviour in response to the change in the relative prices of high- and low-carbon goods. The extent to which these adjustments are made depends on how much the polluting goods and services matter to us – how responsive or elastic demand is compared with price. Some things really matter to us and we will pay almost anything to get them. Others matter less.

When it comes to carbon, nobody particularly wants it. Demand for carbon is indirect – it is the goods and services within which carbon is embedded that count. The carbon price helps here because the companies that strive to meet our demands will want to keep their costs as low as possible. A carbon price means that those inputs with a lot of embedded carbon that a company uses to make things and deliver services will be more costly. Companies will therefore strive to substitute carbon-intensive inputs for non-carbon inputs.

Again responses to the price will vary. Some activities require carbon-intensive fuels, or carbon-intensive steels and cement. Currently there are few alternatives to using fossil fuels for many of these. Others have more viable substitutes. Time matters here too. In the short run, many of the inputs are given. Changing an energy system, for example, takes time. But in the longer term, if there is a long-term carbon price (or the expectation of one) then it makes sense to invest in low-carbon fuels. Which ones depends on relative costs – to each other and the carbon price.

Each technology lobby will claim to be the cheapest, and try to persuade governments to pick them. That is the all-too-pervasive reality under the current command-and-control policies. With a market price, companies have to put their money on the line. So if they are wrong, they will lose money – and vice versa.

It cannot be stressed too strongly how powerful the carbon price is in undermining the carbon pork barrel, and with it all the lobbying and vested interests that exploit government decision-making. The rents from lobbying are removed, since with a carbon price the only thing to lobby about is the level and coverage, compared with, for example, the opportunities to secure a specific subsidy granted by a specific minister or government department, and compulsory market shares. It is therefore not surprising that some of the most expensive technologies have the biggest lobbies and are most opposed to using markets. They have most to lose. The higher up the cost curve the technology is, the greater the lobbying. It is no surprise that offshore wind is such a lobby-intensive activity.

Politicians may also oppose carbon pricing when it exposes the technologies they have supported. Exposing wind projects to competitive bidding against other ways of reducing carbon is often ruled out by specially protected and reserved markets. The Renewables Obligation works in this way in Britain, as do technology-specific FiTs across Europe and the Renewable Portfolio Standards in the US.

Markets are neutral as regards the demand and supply sides: they look for the most efficient responses, and usually this is in both demand and supply. They lock out lobbying and rent-seeking. But, as with any policy instrument, what matters is the design of the market, and in particular the level of the price.

Market design and the price of carbon

There are broadly two ways of fixing the price of carbon. One is directly, via a carbon tax. The other is indirectly, via emissions trading. It turns out that, for rather complicated reasons in the case of climate change, the former is much better than the latter, but as with so many aspects of the climate change challenge, the latter has been chosen, especially in Europe.

In a certain world, it does not much matter. With all the information, the government can either fix the price of carbon, or it can fix the quantity of carbon produced and allow it to be traded to reveal the price. The carbon tax will be roughly equivalent to the price of the carbon permits.[3] But, as noted above, the case for markets rests on uncertainty, and it matters a lot what we are uncertain about, and what limited information we have to describe our degree of uncertainty. It is inevitably messy.

The classic argument for deliberating between carbon taxes and carbon permits depends on what happens if our guesses about the future turn out to be wrong. Suppose, for example, that the price of mercury pollution in a river is fixed on the basis of an estimate of how much pollution factories along its banks might discharge at that fixed price. Now suppose that the estimates turn out to be wrong, and they discharge more into the water than expected. Mercury is very poisonous, so a bit more pollution would have very serious consequences. It therefore matters more in this case that we can be certain about how much mercury they will emit than worrying too much about the price. The damage function is a very steep curve. This would be a case for strict permits, and we should go for tradable permits, or even an outright ban. Taxes would at best be inappropriate, and at worst might lead to dangerous mistakes.

Now consider climate change. Suppose we estimate that a certain amount of carbon will be emitted from our cars, heating systems, power stations and factories. But then suppose again we get our estimates wrong, and a bit more or a bit less is in fact emitted. It will not make much difference to climate change – a few more or less tonnes of carbon are neither here nor there. Now suppose we nevertheless went for permits and fixed the quantities, and it turned out that we had underestimated the costs of reducing emissions. The extra costs might be large, especially in the short run – for example, if new nuclear power stations have cost overruns, or wind turbine costs do not fall as much as some predict. In this case, we should therefore care more about getting the costs wrong than the quantity, at the margin. Hence, to be certain about the costs, we should fix the price, and worry less about the precise level of emissions that will result. Give or take a few tonnes, it doesn't much matter.

These cases illustrate a general policy recommendation: fix the quantity when the slope of the damage function is steep relative to that of the cost function; fix the price when the cost function is steep relative to the damage function.[4] Which one is carbon? As we have seen, the answer is that it is likely to be the latter, and hence a carbon tax is better than a permits scheme. Furthermore, the practicalities of a permit scheme mean that the permits themselves are likely to be short-term – a few years. So the price is short-term too. A carbon tax can be set much further into the future – a feature which turns out to be crucial for the incentives for R&D and new technological developments.

The European Union Emissions Trading Scheme

Why then did Europe end up with emissions trading and the EU ETS, which handed out emissions permits to the major power stations and large industrial plants across the EU, and then encouraged them to trade them, thereby establishing a carbon price? The answer to this question turns on another aspect of setting prices. When a price is put on carbon, people (and companies) are affected in two ways. The price change induces a substitution from high carbon to low carbon, and encourages energy efficiency. This is the good bit. But the price also affects our income (and the profits of firms). Before the price appeared we were not paying for the carbon pollution. Suddenly we are. The price of petrol and fossil fuel energy goes up – and in consequence so do the costs of chemicals, steel and cement. We are worse off. Our shopping basket costs more. This is the income effect which we have already encountered in the discussion of energy efficiency.

The income effect of a carbon price is likely to be large in the short run relative to the substitution effect. We cannot easily and quickly switch to low carbon. Not surprisingly, this is the focus of attention – and politics. Who gets the income, and who pays it, matters a lot. In the carbon tax case, the money typically goes to the government. It is a fiscal transfer. Polluters are worse off. In the permits case, who gets the money depends on who gets the permits and whether they have to pay for them. If the permits are given out – grandfathered – then the money stays with the polluters. They still have an incentive to substitute away from carbon, but the value of the permits remains with the companies. Only if governments auction the permits do governments get the money.

It is immediately clear which way the polluters will want the carbon price to be set. They will want permits rather than taxes, and they will want them to be grandfathered – i.e., based on their prior pollution levels, handed out free of charge. That is exactly what their lobbying achieved. The early attempts at carbon taxes – notably the European Commission's proposal in 1991[5] – were shot down in a barrage of industrial lobbying.[6] It turned out that the EU ETS proved sufficiently polluter-friendly that the transfer of value in the permits generated windfall profits. The companies were given a valuable asset – the right to pollute – for free.

But there was worse. The EU ETS not only proved itself to be capable of generating windfalls for polluters, but also produced low, volatile and short-term prices. The outcome is not what its architects intended, but it is inherent in the scheme's design. The permits fix the quantity of emissions, but that quantity is determined by a wide range of factors, many of which have got nothing to do with carbon per se. Fluctuations in economic output have become the most important determinants of the carbon price. Deindustrialization – driven by relative competitiveness – has also played a part. Finally, the other climate change policies, to the extent that they were successful, reduced the permit prices further and hence encouraged more offsetting coal generation.

These factors drove down the price of carbon in the EU ETS, as the economic crisis took its toll on European industry. The initial 'learning' period – 2005–08 – was followed by the main action in 2008–12. This just happened to coincide with the worst economic downturn for nearly a century. The result was inevitable – and predictable. The price collapsed.

That suited the polluters too. A low price meant that the impact on competition between low- and high-carbon technologies was muted, protecting energy-intensive industries. The silver lining was that the carbon-leakage problem – the competitive threat from imports from countries without a carbon price – was more limited than it might have been.

The volatility that has characterized the EU ETS has two further related consequences. For investors in low-carbon fuels it adds more risk, and this in turn leads to a higher cost of capital if that risk cannot be diversified. But it is great news for financial traders. Volatility stimulates trading, and as a result creates profits for financial institutions. Carbon trading has indeed become an industry on the back of the volatility. New financial instruments have been created to hedge the carbon price, and it comes as little surprise that these financial institutions have started lobbying in support of the EU ETS, which has proved such a profitable opportunity.

The volatility is reinforced by the short-term nature of the periods. If the scheme is reset at regular intervals, attention focuses on two uncertainties: what the permit level and distribution will be for the next period, and how any unused permits in one period will (or will not) be carried over to the next. For the investor in low-carbon technology this adds a whole new layer of political and regulatory risk, as the rules change from period to period. Planning a new technology to deploy after, say, 2020 now becomes a guessing game about the conduct of politicians and regulators over a decade or more.

Short-term political and regulatory risk is particularly damaging to the cost of capital, since it is risk that managers can do nothing about. As already noted, low-carbon technologies tend to be capital-intensive and hence most exposed to the cost of capital. Placing exogenous risk on those unable to mitigate it raises the cost of capital. Risk should be assigned to those best able to manage it – here we have the opposite.

Recent events have done nothing to assuage these concerns. Quite the contrary – as the carbon price plumbed new lows at the end of 2011, politicians started to consider manipulating it by reducing the number of permits. The result would be to turn what was supposed to be a regime in which the quantity was fixed, and the price consequently an outcome of the market process, into a scheme where politicians determined the target price, and then manipulated the quantity of permits to meet it. It would be, in effect, a carbon tax, but with all the associated costs and complexities of the EU ETS. The graph opposite illustrates the sorry recent history of EU ETS prices.

Just when the Europeans were grandstanding again at Durban in December 2011, the carbon price in the EU ETS had fallen to around €6 per tonne. As Johannes Teyssen, Chief Executive of E.ON, stated in

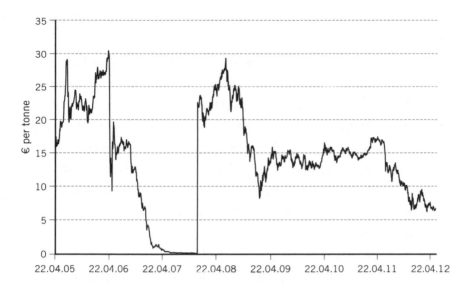

EU ETS forward price history, 2005–12 (€ per tonne)

Source: Bloomberg

February 2012: 'let's talk real: the ETS is bust, it's dead . . . Does the price give any signal for new investments? No. None . . . I don't know a single person in the world that would invest a dime based on ETS signals'.[7]

This was not what its creators intended. Whilst it could be argued that no one had foreseen the economic crisis that would bring down carbon production, the lack of foresight reinforces the argument in favour of the carbon tax.

By 2011, the EU ETS was clearly not doing the job intended, and it provided little signal to investors looking further ahead and evaluating new low-carbon technologies. It added little informational value to longer-term investments, like nuclear, for those countries that might want to encourage its development. The only real impact the short-term price could have was on the relative price of gas- and coal-fired electricity generation – the spread between the two influencing which source of electricity would be produced at the margin from existing power stations. This short-term substitution matters a lot – as we shall see in the next chapter – but even here, few could argue that the EU ETS price reflected the full environmental damage caused by emissions, or was high enough to make much difference.

This, perhaps inevitably, led governments to take another look at the carbon tax that the lobbyists had done so much to undermine. The problem in Europe was that so much political capital had been sunk into the EU ETS project – in order to 'lead the world' towards a global carbon permits market – that it is not likely to be killed off. We appear to be stuck with it. Governments therefore turned instead to the idea of introducing a carbon tax *alongside* a permits scheme (as well as manipulating the number of permits). Thus began a further step towards fixing both the price *and* the quantity.

Floor prices of carbon are a compromise. They stop the price falling below some threshold, and set it for the longer term, and they get over both the problems of volatility and the low price level.[8] A number of countries have started taking this approach.[9] Such carbon floor prices can be made to stick where there is an element of political agreement – for example, that the price floor will never fall. Institutional support can be provided by using arm's-length organizations in the price-setting process, although no government has yet ceded this level of control.

At what level should the carbon price be set?

Setting the 'right' carbon price is no easy task. There are broadly two ways of going about this. The first is the conventional economic approach, using cost–benefit analysis. The basic idea is that an estimate of the marginal damage caused by an additional tonne of carbon emitted into the atmosphere is calculated, reflecting the expected economic impacts of global warming. These impacts of the additional tonne – from sea-level rises, to loss of crops and biodiversity – are all added together, while any benefits are subtracted, and then a marginal value is derived. It is called the social cost of carbon. The calculation should of course be *global*, since climate change is a global phenomenon. The marginal damage in a particular country is largely irrelevant, except where it is sufficiently significant at the global level.

In setting incentives, this marginal damage from emitting an extra tonne of carbon is what matters, and this depends very much on when (rather than where) it is emitted. Thus a tonne of carbon added now when the concentrations are below 400 ppm might have a different effect than an additional tonne in 2020, when concentrations may be above 400 ppm. The social cost of carbon will probably tend to rise over time.

Getting to any precise estimate of the price of carbon is likely to be an impossible task, and yet there have been plenty of attempts. Perhaps the best has been associated with David Pearce's pioneering work. He carefully considered the results of a range of studies, and provided a powerful critique of the official British study that was used to justify preferred policies. Back in the early 2000s, he suggested that the social cost of carbon might be less than £10/tonne, and was probably as low as £3–£6/tonne (compared with the then government's estimate of £70/tonne). His work was reinforced by that of Robert Mendelsohn and Richard Tol.[10] Since the bulk of the world's wealth lies in temperate zones, the costs should be correspondingly lower if we confined ourselves to a narrow GDP measure of welfare. These are but examples of what is an enormous academic literature.

Rightly, these estimates are hedged by all sorts of caveats and health warnings. The range of numbers is correspondingly large because of the time periods involved, the choice of discount rate, and the weights given to the interests of the poor relative to the rich (the so-called equity weighting). There is inevitably an ethical dimension in coming up with such estimates.

The second way of deriving a price starts the other way around, and thereby avoids many of these disputes. It assumes that we do not know what the marginal damage is. The best we can do is have a rough guess. In the short run, there is little that people or companies can do to change matters, so if we set a high price immediately, the income effect will massively dominate the substitution effect – with the inevitable political backlash. There is therefore a good case for setting a low price in the short run, but at the same time creating the expectation that it will be higher for the period ahead over which the capital stock and household appliances may be replaced.

Uncertainty is dealt with by learning from the impact of the initial carbon price – a learning-by-taxing process. A carbon tax is set, and then we wait and see what happens. If the resulting outcome is an insufficient reduction in carbon emissions, the tax can be increased, and vice versa. It is a bit like the problem for central banks setting interest rates to control inflation.

It is made easier if there is a given carbon emissions reduction target already set. In principle, this should itself be the outcome of some attempt to work out the costs and benefits, but as with the above discussion, this may be just too difficult. This is where the 2°C and the science come in. For a variety of both scientific and pragmatic reasons, there is a considerable consensus around a target of limiting global warming to 2°C. The exercises in climate modelling are very rough and ready, with very considerable uncertainties. Yet there is again a degree of consensus around the 450 CO_2e ppm concentration levels for carbon, which may correspond to the 2°C. So the best we can get from the science and the climate modelling are the rough-and-ready figures of 450 CO_2e ppm and 2°C as suitable targets.

Given these targets, the carbon tax can be set with the explicit intention of letting it go to whatever level turns out to be necessary to hit the 450 target. We have to start low because the capital stock is fixed in the short run. But if the tax is made credible, and hence if everyone believes it will be raised (or lowered) to the level necessary to hit the 450 number, then investors can have the assurance that their R&D and capital expenditures will be rewarded by the resulting carbon price.

In this, credibility is everything. The trouble with taxation is that governments tend to be short-lived, and this is a long-term problem. The way to square the circle is to design a set of institutions that embed the tax and the expectations about its future levels. The 450 target could be made a legal requirement. Carbon committees can act like central bankers, setting the carbon tax as bankers fix the interest rate. There can be cross-party agreements.[11] None of these is perfect, but all add some elements of credibility. The result would be very different from the EU ETS and tradable permit regimes. Replacing low, volatile and short-term prices with a tax regime explicitly designed to be long-term would generate the expectation that, starting low, the price of carbon would rise, and the volatility would be addressed head-on through the tax-revising mechanism.

Such an approach lacks precision, but for the very reason that precision is impossible. The policy-maker does not know all the costs and benefits. The market reveals the information in a trial-and-error process.

On what should the price be set – carbon production or carbon consumption?

The aim of a market-based mechanism like the carbon tax is to change the incentives on both the demand and supply sides. In order to be effective, it needs to be targeted as close to the externality as possible. Since it is carbon consumption, not carbon production, which best approximates our environmental footprint, we should pay for the pollution that our consumption causes, regardless of where it is produced.

What is then required is a proxy for this consumption. It is very difficult to devise a tax that exactly targets the carbon embedded in each product we consume. This would be a computational nightmare. Sorting out the carbon in each type of bread would have to distinguish between different ways of baking it. We would need to do the carbon analysis of the supply chain of each and every good and service.

Given this practical impossibility, the carbon tax needs to be targeted at the next-best option, and for this it helps to distinguish between carbon consumed from goods produced domestically, and carbon imports. In terms of domestic carbon emissions, there are quite a few ways we could do this. The obvious approach is to focus on the carbon inputs – coal, oil and gas – and to tax each according to its carbon content. Such an upstream tax would be simple and extremely effective.

An alternative – and the one used in the EU ETS – is to start by taxing those plants that are already covered by regulation, and in particular the EU's LCPD, and then gradually expand the coverage. The plan to include aviation within the EU ETS is an example of a step-by-step approach. This has provoked a vociferous, although not wholly unexpected, backlash from the US, China and India, among others; yet such reactions may lead to an incentive for polluters to take their own carbon measures to avoid the need for permits. We return to this point below.

A gradual expansion of the domain means that, over time, more and more production activities are required to measure their emissions. It focuses minds, but adds bureaucracy and cost, and limits the scope to those that can in fact do the practical measurements.

One reason why the simple fuels-based input approach might also be useful is that some of these fuels are taxed already, and it is inefficient to tax them inconsistently. Thus petrol is taxed, but not coal, and in Britain oil and gas produced in the North Sea are taxed, but coal produced in British mines is not. Indeed, so high is the tax on petrol and diesel in many European countries that it might well be greater than it would have been had the tax been designed purely on a carbon basis. In the US, the adjustment would need to take account of the much lower fuel taxes. There are then a variety of ways of doing the internal bit. The difficult issue is how to deal with carbon imports.

The role of border taxes and border carbon adjustments

If the right domain for the carbon price is consumption and not production, then the question of imports and carbon leakage has to be addressed. As noted, unless carbon leakage is addressed, the US and others are unlikely to join in appropriate action. It is at the heart of the international problem. There is little point in taxing carbon in the home market only for production to shift overseas. In fact, such carbon shifting tends to make matters worse, since the goods and services need to be transported and are often produced in less efficient and more polluting ways. Shipping and aviation are carbon-intensive methods of transport, and the coal-fired power stations and steel mills in China typically have lower thermal efficiency and fewer controls of the various emissions.

How can we address carbon leakage? The easy answer is to tax the carbon content of imports. The hard answer is to explain how exactly this will be achieved. The difficulty of working out the carbon content of particular goods and services was noted above, and it would be practically impossible to disaggregate at the importing port.

Fortunately this isn't necessary, unless we aspire to achieve a utopian carbon-pricing perfection. It is better to be a bit right than exactly wrong, and hence if a carbon price can be applied to some of the most polluting imports, progress will have been made. It turns out that a very small number of industries are responsible for the bulk of carbon imports – steel, chemicals, aluminium, cement, and fertilizers, for example. If we could find a rough-and-ready way of taxing these, then the problem of carbon leakage would be much reduced.

The obvious way to do this is to set up some broad categories and make some assumptions. It is not hard to make an educated guess at the amount of energy used in making steel, and the composition of the electricity industries in the exporting countries is well known. China's electricity generation, for example, is about 80% coal-based. Some broad numbers can be applied to each case. The calculation is affected by the environmental policies in the exporting countries. Where there is a carbon tax in the exporting country, this would form a 'tax credit' to be set against the border tax.

There have, not surprisingly, been a number of objections raised against carbon border taxes and related adjustments for emissions trading systems. The most obvious is the claim that it amounts to protectionism. However, a moment's reflection points towards the opposite conclusion: *not* to have a carbon price is a trade distortion since it represents the subsidizing of polluting exports. Although countries with an enthusiasm for protectionism may use the environment as an excuse, the absence of a carbon price remains a distortion whatever the motive. So the protectionist argument is simply wrong.

Other objections include the claimed incompatibility with WTO rules. This can be dismissed on two grounds: WTO rules allow for environmental considerations; and, more importantly, the WTO rules can, and should, be amended. A further objection is that a border carbon tax would reduce global trade. This is probably correct: taking carbon into account would increase transport costs, and that percentage of trade that arises directly from carbon leakage would have its economics exposed. Producing locally nearer to the final market usually has a smaller carbon footprint, other things being equal. A contraction of world trade may be the inevitable result of taking carbon into account: global warming makes local production more important. This is no bad thing.

The impact of border taxes on globalizing the carbon price

The border carbon tax has one further major plus. The tax credit would provide a very powerful incentive on the carbon exporter to impose its own carbon tax or permit scheme. It would as a result keep the revenue from the tax rather than see it paid to the overseas country. Consider the cases of China, India and the US – the major polluters, all without carbon caps, and none with effective carbon prices. Export-dependent countries would now face relative prices that internalized carbon. They would therefore be incentivized to reduce their carbon inputs in order to maintain competitive advantage. Furthermore, as noted above, they would pay the carbon tax to the foreign government, but would clearly prefer to keep the money themselves.[12] It already happens with car emission standards in the US and Europe: Japanese car exporters have to adjust their own car manufacturing to meet these standards if they want to sell their cars in US or European markets.

This is a pragmatic mechanism for gradually translating bottom-up carbon taxes into an international agreement. Rather than the Kyoto path of a top-down agreement, the border carbon tax provides a bottom-up process by which other countries have an incentive to join in the establishment of a global carbon price. It could start in the EU, with its internal carbon price, and then spread out through the EU's trading partners. The coalition of the willing is not handicapped by the self-interest of the unwilling. Indeed, the unwilling would have an incentive to act in order to keep the money and their export markets. If in due course there is a global agreement with credible binding caps after 2020, a border tax regime would be complementary.

The example of the recent disputes about including aviation in the EU ETS has some implications for countries without emissions trading schemes. In response to the EU insisting on all emissions from flights to and from the EU being covered by permits, the US, India and China could create their own trading regimes. Hence the imposition of an EU ETS requirement on aviation would incentivize them to neutralize the transfer payments from the permits, thereby creating a level playing field. The problem with this approach is its complexity, and the inferiority of permits over taxes, as discussed above. Whilst a common price of carbon would make the import credit straightforward and transparent, separate trading schemes would each produce their own prices as a result of exogenous factors like their own economic growth and business cycles. So it would probably not be enough to have a comparable trading scheme. There would still need to be a price adjustment. The eventual aim is a global carbon price.

The border carbon tax provides a fairer way of distributing the costs. The Europeans would be paying for the carbon emissions caused by both their domestic production and their imports, and hence the combination would mean that they were paying for their full carbon footprint. This would meet the burden-sharing demands that developing countries have repeatedly raised at Kyoto conferences, such as Copenhagen and Durban.

Rather than getting off very lightly as under Kyoto, and disguising the lack of impact through the deindustrialization process, Europe would find that the very industrialized nature of the products it continues to buy and consume would be more fully reflected in the price. That is the reality of the carbon footprint, and in international terms the fairness considerations demand that that price be paid.

With a carbon border price, other countries would be encouraged to have their own carbon taxes, and burdens would follow carbon consumption. Once other countries introduce a carbon price, more would probably join, since the border tax undermines the incentives to free-ride. Gradually, an international approach could emerge. The crucial difference is not only that the carbon border tax does not need an international agreement, but also that the Europeans and others can get on with tackling climate change – for which their responsibility is far greater than the 11% of global emissions that their leaders are so keen to quote – without waiting for a possible Kyoto framework agreement that *might* apply from 2020.

With a carbon price in place, focused on what matters – carbon consumption – the market would then be properly incentivized to start the serious business of decarbonizing. Rather than the short-term, volatile and low prices that the EU ETS has witnessed, a carbon tax starting low, and rising into the medium term, would provide a stable framework for encouraging switching between the fossil fuels, and for R&D in renewables. The next two chapters examine the short-term opportunity to switch from coal to gas – which a proper carbon price would encourage – and the development of the new technologies capable of finally solving the climate change problem.

CHAPTER 10

Making the transition

The overwhelmingly immediate question in climate change is how to stop and then reverse the dash-for-coal, and to do it quickly. This dominates all other immediate climate issues: fail to achieve this and we are in serious trouble. It is patently obvious that we need a transition strategy – in particular, one that heads off the enormous projected expansion of coal-fired power stations, and gets the existing ones closed as fast as possible. Anything like the 1000 GW of new coal plant planned for China and India through to 2030 spells disaster for the climate, as do plans for new coal-fired electricity generation in South Africa, the US, and indeed in Europe too. We do also of course need to get out of oil and, in particular, get out of the imminent massive increase in the production of petrol and diesel cars to meet growing demand from the developing world – and that's not impossible. But even this is less urgent than getting off the coal addiction.

It is very fortunate that there is a *transitionary* option: gas. It is only a transitionary option because gas is a carbon fuel too, although, as discussed, its carbon impact is around half that of coal. It is considerably better than oil too. In the longer term, we will need to get out of gas – unless it turns out to be economic to sequestrate carbon emissions by CCS from gas power stations, and on a sufficient scale. But as an immediate alternative to coal and oil, gas provides an opportunity for a major step reduction in emissions from business-as-usual; it can be achieved quickly; and what makes it particularly attractive is that it would almost certainly be a lot cheaper and on a much bigger scale than the other options. It may also be reasonably secure.

There is no certainty that it will happen on a sufficient scale, but there are two reasons for optimism. First, the unconventional supplies of gas are already here, and likely to increase for the period until new technologies can take up the gauntlet. There can therefore be some confidence that it will play a major part in world energy supplies. Second, if a carbon price is in place, it will significantly alter the coal/gas relative price, and hence further encourage the substitution from coal to gas (and from oil to gas too).

The advantages of gas

Shale gas is a fact, not a forecast, in the US, and it is already a game-changing fact. It has already begun to transform energy markets and its geopolitical consequences have already begun to be felt, not least in the Middle East. The emergence of shale gas has made an enormous dent in the credibility of the claims of the peak oil brigade, although many seem to be in denial.

The shale gas revolution has a particular resonance for the electricity market and for the options for new power stations. Gas-fired power stations have the peculiar economic merit of being comparatively quick to build, and have lower capital costs relative to coal, nuclear and wind. Their fuel costs are higher than wind and nuclear (but not coal), and hence the relative economic attractiveness of gas-fired electricity generation is in the trade-off between capital costs and fuel costs. What matters for their economics, therefore, is the gas price – and the security of supply.

The outlook for gas prices has dramatically changed with the coming of shale gas. Gas prices in the US fell to as low as $2 per MMBtu by April 2012, before edging up slightly. This may prove temporary, and some increase is likely given the production costs and the ability to switch the drilling rigs to shale oil production. Forecasting its future path is rather like picking winners, except in the short term where there are already future prices to draw on. But even at $4, it is still very cheap compared with past prices in the US, or indeed prices anywhere else in the world.

In Europe, prices are currently at least three to four times higher. But even at these higher prices, gas remains competitive against coal for new power stations. European gas prices are held up by the long-term, take-or-pay oil-indexed contracts, which the Russian monopoly Gazprom has imposed on its European customers, and which big utilities, especially in Germany, have signed up to – typically ahead of the shale gas revolution.[1] Lags in developing gas infrastructure and LNG terminals also underpin current prices, as do Gazprom's own problems with meeting demand, given the state of its mature fields, its poor infrastructure and its subsidized domestic customers. This need not, however, be a permanent higher price level, and indeed even Gazprom has realized that it is in its interests to renegotiate in order to avoid losing market share. In Japan, oil indexation is also prevalent, although here too there are signs that it is beginning to weaken. A further weakening in the impact of these contracts could occur if the oil price fell back.

These oil-indexation arrangements date back to the days when gas was a premium fuel – so precious that it needed to be conserved for the petrochemical industry. But over time it is hard to imagine that gas and oil prices will remain so closely linked. The US withdrawal from international gas markets – because it no longer needs to import – has improved global supply conditions as the gas destined for the US from countries like Qatar now has to find markets elsewhere. Remember that the US represents almost 25% of the world economy. The full impact of this on world markets has been disguised by the nuclear disaster in Japan, forcing it to massively increase its gas imports. But this is a temporary phenomenon, and over time the US move to rely on the Americas for its energy sources will have a profound impact on world markets. Shale from the US will lower world prices from where they would have been, and these will feed back to Europe and China. Indeed, this has already begun to happen. The extent is uncertain, but the direction of travel is pretty clear.

This would be enough to change the game for gas around the world, but, as noted, it will be reinforced by major shale gas developments elsewhere. China has both the world's largest shale gas reserves and the political will of an authoritarian regime to exploit them rapidly. In March 2012, it launched its five-year shale gas development plan, which includes supporting subsidies and preferential tax treatment.[2] Argentina has discovered shale deposits in the Vaca Muerta region that may match the main deposits in the US. Poland has significant reserves – indeed, European geology points in a similar direction across the continent.[3] The world is potentially awash with gas. The question is what to do about it, and not simply to ignore it or, as some environmentalists and many nuclear advocates might want, to wish it would go away. Finally, the countries around the EU's borders have lots of shale gas deposits. Algerian resources are staggering, and it is likely that Libya, Egypt and the Ukraine have significant quantities too.[4]

The environmental costs of gas – especially shale gas

Gas is not only already abundant and relatively cheap. It also has numerous environmental advantages over coal in addition to the much lower carbon emissions. Coal emits all sorts of nasty stuff, including NOx and SOx, as already noted. From a power station perspective, significant additional investments – and costs – are required in order to remove these pollutants from the emissions from coal-fired power stations. These increase the costs not only to new coal plants, but also to old ones. Retrofitting technology like FGD is an expensive undertaking.

The carbon benefits of gas over coal should not, however, blind us to the environmental costs of gas. Coal may be dreadful stuff, but gas has its own pollution problems in addition to the carbon emissions. For conventional gas, a major issue is the escape of methane. The climate impact of methane is by volume much greater than that of carbon, although it does not stay in the atmosphere as long. Methane escapes from gas wells and gas pipe-lines, and in poorly maintained and old infrastructures – like that of Russia – these leakages can be very significant. To this are added the leaks from gas-producing wells, including those caused accidentally, and there are the other environmental impacts of the upstream facilities, including LNG plants.

Such problems can in principle be at least partially mitigated. Point sources of methane leakage as described above should not be confused with methane leakages from coal mines, which are very hard to control. The pollution should always be measured relative to coal, our key imme-diate climate change problem. Shale gas technology, however, has its own issues, and these have proved very controversial and have led to sensation-alist lobbying against its deployment, particularly by green NGOs and certain elements of the media.[5] As noted, these problems can be serious and they need regulation. They are concentrated on three aspects of the technology: the use of water in fracking; the use of chemicals in fracking; and additional unconventional issues in respect of the escape of methane.

As explained in Chapter 7, shale gas extraction requires a lot of water. This water is injected with sand and chemicals to break open the rocks at depths, so the gas can flow. The gas comes back up with the water, and hence much of the water can be re-used in a closed cycle. Oil may be mixed in too, and needs to be collected and stored. The availability of water obviously varies, and hence the environmental (and economic) question is: what are the alternative uses for the water resources? This can be considered only on a case-by-case basis, and should therefore be subject to site-specific regulation.

In opening the rocks, it helps to inject not only water and sand, but also chemicals. This is perhaps the most controversial aspect of shale gas production. Chemicals entering the water supplies would be bad news, and since the depth of shale gas deposits and their location in respect of water aquifers varies, again a case-by-case regulatory approach is required. Where the shale deposits are deep (and they tend to be) there are often impervious rocks separating the two. Such separation is often not the case in respect of coal mines, where pollution of aquifers includes heavy metals as well as methane. Indeed, if the same conditions were to apply to coal as to shale gas developments, a great number of coal mines would probably be shut down.

As the gas comes to the surface, some methane escapes. This is the case with all fossil fuel resources. These escapes need to be limited and controlled – and hence regulation is again required. But since methane is valuable – that is what shale gas is – it is not in the interests of the field developer to waste it. So capturing it should not only be a regulatory requirement, but also of some economic value as well. Anti-shale gas campaigners highlight other problems. Fracking can – and has – set off small earth tremors.[6] There is visual intrusion, and the gas has to be collected and distributed, again with visual impacts. Where drilling takes place in sensitive habitats and landscapes, there are obvious concerns, and again a site-by-site approach is required. Such landscape intrusions, however, do need to be put in context: wind farms tend to be in sensitive locations, and whereas shale gas drilling is necessarily temporary, wind farms are more permanent structures.

Taking these considerations into account, there will be sites where licences should not be granted, and there is a need for strong regulatory controls.[7] Opponents of shale gas developments have a good case to make about specific locations, and about the general quality of the regulatory framework. This, however, is distinct from a pre-emptive ban, such as that applied in France. Indeed, in the case of France, it is perhaps of note that nuclear has not been pre-emptively banned because of its waste. The competitive threat to nuclear posed by shale gas may be a motive for its ban.

From coal to gas

In addition to building new gas power stations to meet the enormous demand growth, there are three immediate ways of addressing the coal problem: replacing existing power stations with gas when they come to the end of their economic lives; forcing the early closure of existing coal stations; and using coal stations less often whilst they are on the system.

The first of these options is the subject of intense controversy: environmentalists insist that we should switch straight from coal to current renewables, augmented by a good dose of energy efficiency. They typically argue this without regard to the costs, and to the extent that they consider the costs at all, they argue that oil and gas prices are going to go up for peak oil reasons. Hence gas will be too expensive as well as more polluting than wind farms. For the transition they are likely to be wrong on the first point, and the second is more complicated than it might at first sight appear. Gas is already a much cheaper way of getting down emissions quickly than renewables or nuclear. It would take a very high carbon price to change this. The comparison of the impacts on emissions between gas and renewables depends on whether renewables lead to significantly higher electricity prices, which have the effect of displacing energy-intensive activities to higher-pollution locations overseas. The issue in these cases is the comparison between gas-generated electricity at home versus coal-generated electricity abroad.

There is a lot to be said for the second option: forcing early closure of coal; and where this is not followed at least doing the third: reducing the running hours. This is something the developed countries could contribute, and it would quickly make a significant difference to emissions. The good news is that to a considerable extent they already are. Given muted demand growth, the US and Europe could accelerate the closures of old coal plant, and replace them with gas now, rather than waiting for them to reach the end of their economic lives. This could be achieved by raising the carbon price to a higher level than that witnessed in the EU ETS – as recommended in the previous chapter. It could also be done by regulation, and indeed that is what the LCPD in Europe and environmental standards are pushing along in the US. If both Europe and the US were serious about climate change, the rapid closure of *all* coal stations would be the first call.

Gas has a further advantage in that it provides a transition away from oil in the most stubbornly difficult sector – transport. Gas contributes to transport in four ways: cars and trucks can be run on gas rather than petrol or diesel; gas can power hybrid vehicles; it can be used to generate electricity for electric vehicles; and it can be converted into oil products to be used in conventional car engines (gas-to-liquids).

The direct substitution of gas for oil in transport has been a niche market for some time. That is now changing, particularly in the US with LNG facilities for the conversion of trucks.

Hybrids are essentially an intermediate technology – they blend fossil fuels with electricity to reduce emissions by reducing fuel consumption and substituting at the margin from petrol/diesel to gas. The way they work is to use electric power when the batteries are charged up, and conventional fuel when they are not, especially for longer journeys. When fuel is used, the battery is charged and even the heat energy from braking can be captured. They have lots of immediate advantages over pure electric vehicles. They do not suffer from the limitation of range (which current batteries do), and they require less in the way of new supporting infrastructure. Not surprisingly, hybrids tend to be cheaper than all-electric vehicles.

Pure all-electric vehicles hold out the prospect of a significant substitution away from oil. If electric cars really take off, then so will the demand for electricity, and here again the advantage of gas over renewables in the transitional period is very considerable. Finally, gas prices are so low in the US relative to oil that gas-to-liquids is now a possible option. Thus gas can go straight into a conventional car; it can make petrol for a conventional car; and it can make electricity for an electric car. This is why gas and oil can be substitutes – a point that holds out the prospect of lower emissions (the focus here) and undermines peak oil (the focus of Chapter 7).

Gas, CCS and the long run

Gas may play this important transitional role, but there is no escaping its carbon content. Therefore it cannot be a long-run solution to energy supplies, unless its carbon can be captured and stored through CCS technology. Gas with CCS offers a much longer time horizon. But will it work?

The idea behind CCS is a simple one. Just as photosynthesis sequestered carbon throughout the geological ages, there is the possibility that the carbon generated by gas-fired power stations (and indeed from coal-fired power stations and industrial processes) could be put back into the earth from whence it came. We take the energy we need, and return the carbon we do not need back into the ground.

In theory, the CCS route is straightforward.[8] It requires three main steps: the carbon needs to be separated out; it needs to be piped away to a sink; and the sink needs to be securely capped. It is the exact reverse of what happens now: the gas is extracted from a gas reserve; it is transported to the power station; and then it is burnt.

There is little new basic science required to get going on CCS (although there is still a lot of R&D that can contribute and might bring further breakthroughs). Gas separation has been possible for more than a century. Pipeline transmission of gas has been around for a long time. Indeed, coal gas was piped around London above ground in the first decades of the nineteenth century. Many countries already use depleted gas wells and other cavities as storage. The problem is all about the practicalities and the costs.

There are two approaches to separation: it can be before combustion of the gas in the power station, or afterwards. What is required is the fitting of the equipment to power stations to do this. Building the CCS equipment into a new power station is much more straightforward than retrofitting, although in principle both can be done.

When it comes to the costs, it is important to recognize that these have *system* properties. The cost of each element of CCS within a system of pipelines and stores designed with this purpose in mind is likely to be different (and less) than those of a stand-alone plant. But first the technology has to be demonstrated, and there are as yet only a handful of small-scale experiments. It is therefore too early to tell what the costs may ultimately be, and whether gas plus CCS can compete with other low-carbon technologies like nuclear and renewables. This calculation is further complicated by the possibility that there may be considerable advances in these other technologies whilst CCS is being tested. Electricity storage might develop too, changing the costs of intermittent renewables, and a new generation of nuclear power stations might be commissioned.

The fact that it is too early to tell is important in thinking about the design of energy policy. Some suggest that we should now mandate that gas power stations must, at some future date, be fitted with CCS.[9] But such a regulatory control might limit gas investment now (since it raises costs now), and hence effectively close off the gas option for the transition. Indeed some might regard this as a way of killing it off. A better position would be to require that *new* gas-fired power stations be designed such that, if practical and necessary, CCS could be added later, and to leave open the CCS option until we see the lie of the (technological) land in the middle of the next decade.

Even if CCS proves successful for gas power stations, its role in a low-carbon world may still be limited. Whilst the North Sea has many depleted gas fields, this is not necessarily the case elsewhere. Shale gas extraction does not necessarily leave large easily accessible empty holes behind. That is not the way shale gas extraction works. Lack of storage space may be a particularly significant problem for China and India, although there are CCS technologies that inject the CO_2 into coal seams. The coal absorbs the CO_2, releasing methane, which in turn can be extracted in a process called enhanced coal-bed methane.[10]

It is probably wise to conclude that CCS is likely to be at best a limited and local option, and in the medium term gas will probably have to be phased out if the ambitious 2050 decarbonization targets are to be achieved, save for a limited reserve capacity role. R&D on CCS should be encouraged, and if it turns out to be a runner, then that will be a bonus. But in the meantime we should continue to treat gas as a *transitional* fuel, and CCS should for policy purposes be assumed to be a limited offsetting technology option.

If CCS works for gas, it might also work for coal. Indeed, in many cases CCS is mandated for new coal even though the technology has not been tested. It might then be concluded that if CCS works, the gas–coal substitution case falls down. This is not convincing, however. Over the next decade, most (indeed, almost all) coal and gas will not be subject to CCS. This is the period over which the Chinese and Indian economies may double in size, and there is no sign that all of their new coal power stations will have CCS fitted. The big difference in emissions matters – as does the differing thermal efficiencies and the other pollution externalities. Coal is still dreadful stuff even with CCS. There is also the further consideration of the stores: coal does not come out of reservoir-type stores, unlike conventional gas. Finding sufficient sinks for coal-related carbon is likely to be more difficult. Gas CCS has yet to prove that it is economic: coal CCS currently looks more like wishful thinking than a practical proposition.

This disappointing result holds too for underground coal gasification. The idea is a neat one: instead of mining the coal, and transporting it to power stations, it is gasified in situ, and the carbon is left behind. However, not only is there the volume issue, but practical questions arise about not allowing it to escape. Underground coal gasification may turn out to be better than conventional mining but, as with gas CCS, its contribution is to mitigate rather than solve the carbon problems.

The gas option at the global level

In Europe, the gas option can be compared directly with the alternatives – nuclear and current renewables, in a context where new coal is now very limited (except these days in Germany, ironically). Here the option becomes even more attractive if the gas price falls towards the levels experienced in the US at the end of 2011 and into 2012 – which, as noted, have been less than one quarter of the European level. Falling gas prices would not only facilitate cheaper options for carbon reductions, but in the process also free up resources to devote to other aspects of the climate change problem and its mitigation – notably R&D, as will be discussed in the next chapter.

But what matters is the arithmetic of emissions, and emissions growth is not (on the whole) what happens in Europe. It is a relatively small player in terms of global carbon *production* – at around 11%. It is the emissions growth in the rapidly developing countries, and to an extent in the US too, that matters more. Here the gas–coal substitution is really all that stands in the way of emissions growth in the next decade.

The prospects are pretty good in the case of the US, which also has lots of coal, but can switch from coal to gas at little cost, especially when it comes to new power stations. This matters, since the US, like much of Europe, is entering into a replacement cycle following on from the great expansion of coal-generating capacity in the 1960s and 1970s. In the US (and Europe) many coal-fired power stations are old and coming up for replacement anyway. The realistic choice for much of this replacement investment is new coal or new gas: and cheap gas means that choosing gas has no net additional cost. It would be an even better bet if there were a carbon price. The good news is that the substitution has already begun: coal's share of US electricity generation is falling, and the gas share is rising in the US and Europe. The US's carbon emissions are, as a result, falling very fast.[11] But it is not nearly quick enough. In the EIA 2012 forecasts, coal's share of the electricity generation mix drops only from 45% to 39% between 2010 and 2035. 33 GW of coal capacity are retired, but 14 GW of new coal capacity already under construction are completed.[12] Yet these figures might turn out to be too pessimistic and it is important to recognize that the coal-to-gas switch is already having a big impact on US carbon emissions. According to the IEA, in 2011 they fell by 1.7%, and by 7.7% since 2006 – the largest reduction of all countries or regions.[13]

China is the most important and difficult case. To date, it has relied on coal and, as we have seen, it is its coal-burn that has been the major direct cause of emissions increases since 1990. It is 80% coal-dependent for electricity generation, and it is the country where car ownership is likely to expand dramatically in absolute terms. To date its strategy has been to diversify into nuclear and a limited amount of wind. But again shale gas has opened up new horizons. It is early days, and China has yet to develop a shale gas capability along the lines of the US. But the resources are there, and it has ambitious plans to exploit them. If gas can be a substitute for coal at the margin in China, reducing the new coal build, this will be of global significance. China also has potential for other non-conventional gases. Coal-bed methane might have a significant prospect in China, and given the enormous deposits of coal, this could provide some relief from the unconstrained coal – and at the same time capture some of the methane that leaks out of coal mines.

The opportunities for coal-to-gas substitution on the basis of indigenous gas supplies – conventional or otherwise – are more uncertain elsewhere, but as exploration for shale gas progresses, there may be yet more opportunities.

A dash-for-gas and its consequences

Gas is the only serious large-scale competitor to coal over the coming decade. Only gas can make serious inroads into the coal-burn in the US, make a contribution in China, and even have some impact in India. In Europe, it can quickly replace much of the coal-generating capacity. In aggregate, the substitution from coal to gas in the US may make it the main achiever in emissions reductions over this short-to-medium-term period.

Yet in recognizing the lower carbon emissions of gas compared with coal, we should not lose sight of the downside of another 'dash-for-gas'. There are the usual (but often misguided) concerns about security of supply and vulnerability to price shocks. But there are other good reasons why many environmentalists take a more holistic view. It is not just the purist argument that we should be getting out of all carbon-based fuels immediately. There are other more pragmatic concerns: about the environmental impacts of extracting the gas; lock-in; and about the impact on the development of future renewables.

There is no escaping the environmental impact of both natural and shale gas production. As noted, methane may be short-lived in the atmosphere, but it is potent, and many gas pipelines – notably in Russia – leak a lot. Where shale gas is concerned we have examined the further environmental objections. These are real and important. They need proper regulation, and the full costs of shale gas production need to be identified. If this is done properly, some shale gas production may be uneconomic. But there is a great deal of difference between proper regulation and outright bans. The latter assumes that the costs exceed the benefits, and such bans are not without consequences. If shale gas were to be banned worldwide, more coal would almost certainly be burnt, and emissions would rise accordingly. This trade-off is already being seen in Europe. With the immediate closure of some of Germany's nuclear capacity, and its complete phase-out within a decade, there is a stark choice: build new gas or build new coal generation – because current renewables cannot bridge the whole gap over the coming decade. The choice they are making is bad news for the climate. If shale gas were to be banned globally then the price of natural gas would be higher since total production of gas would be confined to natural gas. Supplies would be less secure, and coal's attraction would be all the greater. It is understandable that many greens do not want to get their hands dirty with gas, but they have to face up to the alternative: more coal, more emissions, and more climate change.

A second objection to a gas-for-coal substitution is that it would create lock-in. The gas power stations built now will be on the system; their capital costs will be depreciated, and hence they will have an economic advantage, not just for the next decade or so, but for the medium term. We will therefore be locked into gas until at least 2030, and perhaps beyond. The supporting infrastructure – pipelines, shipping, LNG terminals and storage facilities – will be built, and therefore there will be a strong incentive to go on using it.

This is a very serious objection, and there are two possible responses. The first is to avoid lock-in by not choosing the coal-to-gas transition at all – if gas power stations and infrastructure are not built, they cannot be used. The problem with this approach is that coal generation will probably be built instead, and that too will be locked in, and this would be a lot worse environmentally.

A second response is to explicitly recognize up front the lock-in problem, and to build into energy policy *now* a rising carbon price. If the proposals for a carbon consumption tax were to be adopted, and the carbon price allowed to go to whatever level is necessary to achieve the decarbonization trajectory through to 2050, then more gas on the system would force a rising carbon price. As other technologies materialize, this carbon differential would bite and investors would develop both the generating capacity and the infrastructure with this time horizon in mind. With the prospect of a carbon price that in the medium term limits the contribution of gas, investors would have a greater incentive to consider CCS, both in designing reversible gas infrastructure where carbon could be pumped back into the empty cavities, and also in considering how CCS might be retrospectively fitted. They would also have to account for the possibility that gas plants would be phased out towards the end of the next decade.

The third objection to a dash-for-gas is that it might undermine the economics of current renewables and new nuclear generation. Indeed, there can be little doubt that it already does this, and for good reasons. Coal-to-gas is likely to be a much cheaper way now of reducing emissions, and hence of meeting overall carbon targets in the next couple of decades. Furthermore, gas generation is less capital-intensive than either nuclear or renewables (both of which are overwhelmingly fixed-cost). The implication is that the subsidies to renewables and nuclear will have to be greater, and this puts the onus on their advocates to justify the higher relative costs. Since they are not the cheapest ways of reducing the emissions, the justification shifts more towards supporting immature technologies and R&D, rather than wind farms and rooftop solar.

The obstacles to a rapid gas-for-coal strategy

The good news on the gas transition front is that since the relative prices of coal and gas, before taking account of their pollution costs, are quite close, and since quite a lot of old coal stations need replacing anyway, a move away from coal to gas should be at negligible cost. Add in a carbon tax, and the market should in principle be able to manage the transition on its own. This is a world away from the intensive intervention needed for renewables and nuclear.

Unfortunately the reality is not quite so simple. A number of obstacles stand in the way in Europe. As noted in Chapter 4, the promotion of renewables such as wind brings a lot of zero-marginal-cost electricity generation onto the system. Thus when the wind blows, everything else is forced off the system. This is made even worse when there is regulation to give wind priority for dispatch onto the electricity grids over other technologies. The result is to make new gas power stations dependent for their profits on the amount of wind on the system, and in particular it makes these gas stations intermittent too. They cannot be sure that they will run.

Thus, in order to ensure that enough gas stations are built, they need to have long-term contracts for the capacity they bring onto the system, to mirror those already given to wind. In other words, the coming of wind on a significant scale requires a reform of the electricity markets, not just to protect the interests of the wind generators, but also to ensure that other technologies are not placed at a contractual disadvantage. Governments tend to think they can tinker with one technology without creating consequences for others. They are quite wrong. Once you start interfering in one area of the electricity market, you are forced to interfere in other areas too.

This is necessarily a technical issue, and a very complex one at that. But it is of the utmost importance: if the electricity market is not reformed to take account of the distortions that policy-induced wind energy causes, the result will be fewer and more expensive gas stations, and that in turn will further reduce security of supply and competitiveness, as well as enabling the coal-fired power stations to drag out their lives on the system yet further.

For the reality of the situation is not gas versus renewables instead of coal, but rather gas versus current renewables *plus* old coal, with all the additional carbon emissions that keeping the coal burning will bring.

Across Europe, this is indeed what is happening. Coal is getting a life raft on the back of the renewables. It is further spared more immediate closure because, as we have seen, the renewables targets undermine the EU ETS carbon price, and hence protect the coal power stations from the full force of an appropriate carbon price. As a result, the main factor limiting coal is the regulations on its other – non-carbon – emissions, rather than the carbon price.

Creating a longer-term contracting market for all sources of electricity generation is the challenge, and preventing it from being captured by the current renewables and nuclear interests is important. The playing field needs to be levelled, net of the carbon price.

Reforming the electricity market is one step in facilitating the gas-for-coal substitution. In transport, the gas-for-oil substitution can also be market-led, although there are important infrastructure issues that are subject to government decisions. Then there is the design of the taxation regime for the various fuels for cars and trucks. Finally, there are standards and regulations. Government can do much here, provided that it is clear about its overarching carbon objectives.

These policy measures are all important, but they will make little difference if governments – like the French – stop gas in its tracks. This is obviously a big obstacle, and one that many green groups have campaigned for across Europe and indeed in the US too. But unless gas is used to get out of coal as fast as possible, global emissions will just keep going up, and within a decade or so it will be too late to stop significant global warming. In the process gas will displace some of the renewables and nuclear programmes *now*, but not in the future. Building lots of offshore wind farms *now* is a very expensive way of making a relatively small contribution to emissions reductions today. But in the future, as the carbon price rises, new renewables will play the main part, as, just possibly, will nuclear. It would be nice to be able to afford *all* these options now, but many green NGOs are reluctant to face the reality that financial resources are limited. This is ironic, since scarce resources and their allocation should be at the heart of any green argument.

Investment now in renewables (and nuclear) is money not spent on other options. If gas displaces some renewables and some nuclear generation now, it leaves money to be spent elsewhere. Since none of the existing technologies can achieve decarbonization, new technologies are going to be needed – and the R&D will therefore have to be financed. Better to spend now on that R&D some of the money saved by a less aggressive renewables build, whilst at the same time directly attacking the dash-for-coal that threatens to create a climate disaster in the short term. It is to this technological question that we now turn.

CHAPTER 11

Investing in new technologies

None of the existing technologies, save perhaps nuclear, has the capacity to provide a substantive impact on emissions sufficient to make decarbonization a realizable objective. Nuclear has its own problems, and there is little or no possibility that it will be deployed on a very large scale any time soon. Existing technologies have much to offer, but are nowhere near up to the task. Gas is a *transitional* option that ought to be taken seriously. But beyond that, the gap between what needs to be done and what is on offer is just too great. The carbon problem will not be cracked with what is currently available.

An inescapable conclusion follows: without significant technological development of both existing low-carbon technologies and new ones, climate change is unlikely to be limited to anything like 2°C. There is therefore little choice: new technologies are needed and so R&D has to be a major element of our climate change strategy. It is not an option, but rather a necessity.

At first glance the omens do not look good. Over the last century there has been surprisingly little technical progress in energy supply. Some of the electrical cables in London are Victorian. The conventional coal power station boilers that still dominate the world's electricity systems were first deployed in the late nineteenth century. Gas turbines came from Second

World War developments, and nuclear reactors are largely based on 1950s designs. Electricity is still produced largely by burning oil, gas or coal, or as a result of a nuclear reaction. Hydroelectric schemes – as near as one can get to a zero-carbon technology – have also been around for a long time. But that, for electricity, has historically been about it. In transport, trains have moved from steam to diesel, and now increasingly to electricity, and cars have used the internal combustion engine for a century (although the concept of an electric car is also nineteenth-century in origin).

The reasons for this relative lack of basic technical progress are many and varied. Yet there is one overwhelmingly powerful factor: cheap fossil fuels have blunted the need for change. If oil and coal are plentiful and cheap, why fix what isn't broken? The real price of oil remained pretty stable for much of the twentieth century, as did the price of coal. In the latter case, almost all of it was domestically produced and hence for most countries was a secure source of supply.

Two things have changed: the prices of fossil fuels have risen and environmental regulation has increased the costs of pollution. Although fossil fuel prices are still modest, and there are good reasons to think they may remain subdued, if the world is to address climate change this will have to be addressed by influencing the price through taxing or permitting carbon. That is necessary but not sufficient. It will need to be augmented by policies that promote R&D.

Despite this very static and stable technological picture, there is now an enormous R&D effort under way across universities, company R&D departments, and in specialist institutions. Already there have been results, as drilling techniques (horizontally), seismic information technology and fracking have transformed the fossil fuel industries. The new technological opportunities are abundant and, by definition, we do not know which ones will prove economically and environmentally successful. Many will be dead ends, and much of the R&D spending will inevitably be wasted.

The new and emerging technologies may enable us to address the climate change problem, provided that coal has not wreaked too much damage in the meantime. The trick is to get from here to there, given that R&D will not happen spontaneously in the required time frame. This is where the bulk of the money should go: inventing the industries of the future. That is the policy – and political – imperative.

Our task is not to predict which specific technologies will be the winners. Picking winners is typically not a good way to go. Yet despite this we are not completely in the dark. Although there will be lots of surprises, the sorts of areas where breakthroughs might come can already be sketched, as can the ways in which they might turn into game-changers. Very exciting things are going on, with a good chance of uncovering not only new energy supplies, but even revolutionary changes in the way energy is generated and consumed. So although we cannot pick the winners, we can pick out the rich seams for research.

The first of these rich seams is to build on what we have already got – to improve existing technologies. Although each technology has its own natural constraints, there is virtually no technology that cannot be improved. Much more will obviously be needed, and there are several rich seams of more radical technological change that stand out: the convergence of the communications and energy industries; storage and batteries; the electrification of transport; and new electricity-generation technologies. This chapter gives mere a flavour of what might be in store over the coming decades – enough to excite the imagination, and enough to motivate serious public and private investment. It is neither exhaustive nor predictive. Lots of new developments are left out. It is in the nature of R&D to surprise.

Improving existing technologies

Most technological progress is incremental – a gradual improvement on existing techniques. This low-key process makes wind turbines more effective; heightens the thermal efficiencies of coal and gas power stations; improves the depletion rates of oil and gas fields; and refines the safety and efficiencies of nuclear power stations. Transmission and distribution cabling improvements, transformers, and a host of ancillary services are also subject to this process.

The difference between these sorts of advances and new technologies is that incremental improvements are always constrained by the basic physical characteristics of the technology to which they are applied. A wind turbine has inherent limits to its load factor and the size of its blades; a coal power station has limits to the thermal efficiency which comes from boiler technology, and so on.

What incremental improvements can we expect? Perhaps the greatest attention is being paid to current renewables. This is not surprising, given their high cost and the impact on consumer bills. Renewables involve an array of technologies, but here we concentrate on wind and solar.

In the case of wind, perhaps the greatest technical challenges relate to the development of materials and turbines designed to operate offshore in hostile salt-water conditions. Cost improvements come from the organization of development and maintenance and the logistics, but the technology challenges relate to the blade design and dealing with high and low winds, in order to raise the load factor. Then there is the conversion of wind via the turbines into energy. The good news here is that there appears to be considerable potential. A wide range of designs is being considered, and installed wind farms effectively become real-life experiments that provide researchers with important data. The bad news is that, even if these improvements reduce costs by the 30% target by 2020 for offshore wind in the North Sea and the Irish Sea, they still may not be able to compete with the alternatives.[1]

There are technical reasons for this. Turbines are inherently small: there is a limit to what can be mounted on a mast, on land or at sea. Even if the size of the turbine is bigger – say 5 MW, or even 7 MW – it still might take up to 200 turbines to have the *intermittent* capacity of a modern baseload gas- or coal-fired power station. Even this, however, overstates wind: the load factor is much lower. Even if it rises to 50%, there is still the other 50% to make up. A load factor on a baseload modern CCGT can be around 90% by comparison. In addition, the bigger the turbines, the further apart they must be, and the greater the area of land or seabed required. This is the point made in Chapter 4: wind simply cannot add up to more than a marginal contribution to decarbonization. Even a breakthrough in storage – to which we turn below – would not solve the scale issue.

Progress on current solar technologies is also limited by scale. Whilst it is possible to envisage large-scale concentrated solar thermal plants in areas with lots of sun and lots of land – like North Africa, the Middle East and Central America – the arithmetic of the area of solar panels required is daunting. As a result, most solar is small-scale and local. Technology improves the ability to convert sunlight into usable energy and, on a much more rapid scale than for wind, new materials are making a considerable difference to both efficiency and costs. But as with wind, current solar remains essentially marginal without a major technological advance. Incremental improvements to solar promise much, costs have already come down a great deal, and it may deliver eventually, but probably not in the next decade. It is likely to be a medium-term solution, based on new solar technologies – but it could be an important one, especially in places like India, where decentralized power has a lot to offer a country which has many areas yet to benefit from electricity.

Nuclear differs from renewables in that it is large-scale and baseload. It does not suffer from intermittency, and hence does not typically require back-up. Yet its contribution is limited by politics. Some nuclear enthusiasts hold out for considerable advances, and there are all sorts of new reactor designs being promoted. These should be treated as new technologies. Here the question is whether the existing designs – notably the PWR – can be significantly improved.

As with wind and current solar, the scope for technological progress is limited by the inherent characteristics of the current vintage of nuclear reactors. These mandate large power stations, high levels of safety measures, and lots of ancillary precautionary investments in case of failure, including back-up power stations, multiple cooling mechanisms, and so on. As the accident at Fukushima demonstrated, even these can fail.

Improvement means reducing the costs so that nuclear can compete. The evidence here is not encouraging. We noted that the new French design of PWRs being built in Finland and France has so far proved remarkably resistant to cost control. Both projects are well over budget and well beyond the original timescales. The reasons for this relate to technology and risk. Regulations push the safety boundaries and technology has to respond. Each generation of nuclear claims ever-greater safety in design and, not surprisingly, the costs remain stubbornly high as a result.

The final electricity-generation case for incremental improvements is conventional fossil fuel generation. The coal boiler is, as noted, an old design. Technical developments focus on the temperatures for combustion, including so-called super-critical power stations. There are new ways of preparing the fuel, including gasification. These improvements are often referred to under the generic title of 'clean coal' (although calling coal 'clean' is questionable). Technologies can get better at capturing pollutants, and research is currently focusing on CCS.

CCS is similarly being developed for gas power stations, and there are incremental improvements in the design of these stations. But again the turbines in these power stations have serious limitations. Improvements are being sought in fuel supply, with perhaps the most interesting being in the gasification of a number of potential fuel sources. These include coal, but also a number of potential waste streams such as methane from landfill and a variety of waste products.

Moving to transport, there has been continuous evolution of the basic technologies relating to motor vehicles. As standards have tightened, engine manufacturers and car designers have made major improvements. There is every reason to assume that these will continue, but they will always be limited by the constraints of the internal combustion engine. Indeed, whilst engine efficiency has improved a lot in the last two decades recent improvements to fuel efficiency have come as much from reducing the size and weight of cars and reducing the engine size as from making car engines more efficient. The internal combustion engine has lasted a century – it is as old as the boilers in coal power stations.

It would be a mistake to rule out advances in any of these technologies, just as it would be a mistake to concentrate on the gains in one without recognizing the potential for advancements in the others. Wind and current solar may advance, but so too might competing fossil fuels. Indeed, it was the failure to predict the revolution that fracking would bring to the gas industry that has led to further questioning of the contribution from these renewables and nuclear, and it has undermined their economics. A similar pessimism is often the response to predictions of radical breakthroughs for the internal combustion engine. The implication is that whilst each of these technologies will probably advance, and whilst each has a contribution to make, none can offset the immediate need for a fast transitional fuel source to get the carbon emissions down – swiftly followed by new technologies. Here things get very exciting. What follows is merely a possible glimpse at what might be out there.

Convergence of communications and energy technologies

The first major new technological opportunity comes from outside the energy sector. In the last couple of decades, information technology has transformed the telecommunications and broadcasting industries. When I wrote my doctoral thesis in the early 1980s, I had to rely on a typewriter, carbon paper and Tipp-Ex; now there is the internet, smartphones and tablets. In the early 1980s, the fax machine was yet to be made widely available. Now it is largely regarded as a historical curiosity. Those who use and benefit from these new technologies take them for granted, and very few people have any idea what it was like to rely on fixed-line phones, telegrams and an army of typists.

Looking back over the last three decades, communications have been transformed, and with them the economy itself, our households, and even our social lives. Only the motor car, electricity, and before them the railways, were such transforming general-purpose technologies. What did people do before Facebook and Twitter? How did they keep in touch without mobile phones?

Remarkable as it might seem, even now these new information and data technologies have made no real difference to the energy or transport sectors. Whilst it is true that modern energy markets would be hard to run without computers, and customers might not have been able to switch suppliers so readily, the core systems in gas and electricity remain much as they were two or three decades ago. Today's power stations and the transmission and distribution systems would be recognizable to an electrical engineer in the 1940s and 1950s. Similarly on railway networks, signalling is largely as it was decades ago, with lots of manual interfaces. On roads, advanced traffic management and smart tolling systems are the exception rather than the rule, and deployed mainly in large cities. Traffic management is still largely about queuing.

All this may be about to change for energy and transport, with radical implications for the markets and systems of both sectors. Information technology may even change the structure of the industries, which have traditionally been dominated by large integrated utilities, just as it did to the telecommunications industry. Electricity lends itself to information technology because it is data-rich, and because it needs a network. Coordination is of the essence, especially in the absence of storage, making it essential to match supply and demand instantaneously. This problem has always been there, but what makes it more relevant now is the introduction to the networks of new forms of generation, notably decentralized, intermittent generation from wind and solar.

The absence of active information technology has meant that electricity systems have developed as largely *passive* systems, using overriding command-and-control, typically organized through the grid and the system operator. Without storage, and with no ability to flex the demand side of the market, the job of the electricity companies has been to build an excess margin of generation to meet surges in demand – for example, at half-time in a major football match when everyone puts the kettle on. This calls for planning and large-scale power stations on the one hand, and on the other leaves no room for flexibility for consumers and their appliances. In our example, if all the fridges and phone chargers were temporarily taken off the system at half-time, there would be a counterbalancing of the system and expensive peak plant would not be needed. At present, there is no way to make this happen.

It turns out that there are many more significant opportunities than this trivial example, but the general problem is the same – how to manage demand over a system to take into account all the variables, from intermittent wind and solar to how we watch television. The coming of new information technologies provides the opportunity to make electricity an *active* system, thereby allowing the demand side of the market to come into play. What makes this possible is smart technology: smart meters and smart grids. It enables measurement to be combined with an element of control.

Smart technology is a way of gathering and processing large amounts of data, and designing systems to use that data to effectively manage their operation – i.e., to add artificial intelligence. It provides for fine-grained measurement of the transmission, distribution and use of electricity. A smart meter measures not only real-time consumption, but also the components of that consumption: the role of boilers and heating systems and household appliances, as well as all the devices we need to charge. This data is very revealing – indeed so revealing that there are serious concerns about privacy. By monitoring your household consumption patterns, it will be possible to find out quite a lot about your routines and even your intimate secrets, and the data will be invaluable in working out the best time to burgle your house (and even if it is worth it).

For the politician, the attraction is that it might allow the consumer (and voter) to play an active role. They can see what is going on and adjust their demands to the price. It gets them involved: they can do their bit, and they can take control of their costs. And they can use electricity more efficiently. But whilst this has a part to play, it is essentially a misunderstanding of both the role of smart technology in the system, and of what consumers may want.

For the system, the data generated by smart meters provides the basis for a much more detailed *control*. But it also opens up the more radical possibility that the system operator might use the smart meters to manage demand on the system. When there are capacity shortages – for example, when the wind is not blowing, or blowing too hard – it would be enormously helpful to be able to manage down business and household demand by remotely turning down heating appliances, reducing power to fridges and so on, especially if these adjustments were each small enough to pass unnoticed, but in aggregate big enough to make a dent in the total electricity demand.

From the consumer perspective, few want to spend their evenings watching the smart meter and rushing around turning appliances on and off as the price varies. Most would probably prefer this to be done for them – i.e., they remain passive whilst someone else is actively managing them. It is a bit like consumer attitudes to cars. When cars often broke down, people would spend their weekends fixing them. Now much of this is done for them through electronics. The car's computer tells you when it needs a service or what is wrong with it, and indeed in some cases it can even be fixed remotely. Weekends are now free for other much more interesting things, and car parts shops have all but disappeared from the high street.

Smart technology brings in the demand side of electricity and heating.[2] But it can also revolutionize transport, and therefore its emissions. Transport shares with electricity the problem of system coordination. Its networks are empty much of the time, and then congested at rush hours. It requires a network of fuel stations for the roads to function, and stations and signals for the railways. Although the pollution is caused primarily by the petrol and diesel engines, it can be exacerbated by the way the car is driven, and how the congestion is managed.

Information technology is gradually reducing the role of the driver and moving towards more active management of the car and its travel. Already, satellite navigation systems (sat nav) have provided the ability to download information to the car, whilst CCTV and other monitoring systems provide real-time (smart) traffic and road management information. In addition, the driver is provided with lots of information about the car itself.

The next stage may be to move towards the driverless car – a world in which the car processes data to understand its situation and travel between points. Cars 'learn' about roads as they travel along them, and indeed so sophisticated are such systems that they could even set the insurance terms in real time, since more repeat journeys increase the car's knowledge and hence decrease the risks. Completely driverless cars are of course some way off, but the technology has considerable potential to manage fuel use and hence fuel efficiency, which in turn may reduce emissions.[3]

In rail, the interesting question is not so much the one presented by governments and the industry – that networks are congested and need to be enhanced – but rather why railway lines are empty for such long periods. Wait at any mainline railway station and the immediate impression is the *absence* of trains. There are two reasons for this. The lack of smart signalling means that trains have to be far apart to prevent accidents; and the lack of sufficient platforms and platform management at any given station means that trains have to be well spaced to avoid collisions and congestion. Information technology can change the former and even help with the latter. Witness urban underground and metro systems. Here lots of trains are only minutes apart.

As with electricity, what the new technologies offer transport is the possibility of much more coordination and active system management. From a carbon perspective, given that the demand for transport is very unlikely to fall, and that it will take time to move away from its oil dependency, opportunities arise to limit emissions by increasing efficiency – both in the use of engines and road and railway systems, and by encouraging more public transport. As with active electricity demand management generally, information technology may also facilitate big increases in energy efficiency in transport.

Storage and batteries

The second promising technological development is in the storage of energy. Imagine a world in which intermittent wind energy could be stored and used when needed. Imagine a world where car batteries are small, compact, and can power the vehicle for long distances before they need recharging. Imagine electric cars synchronizing their charging with the best wind flows. You get the picture: storage is another key pillar to transforming our energy systems.

Put another way, the design of all modern electricity systems is based largely on the *absence* of storage. The systems require excess capacity margins to ensure that peaks in demand will be met. Where more and more wind comes onto the system, the intermittency requires additional back-up capacity. At the limit, a 100% wind system would need a 100% back-up system, doubling the capacity to meet any expected demand.

So what are the prospects of a breakthrough in battery technology in the next 20 years? There are lots of promising options. Some of these are large-scale, including forms of pump storage. Others include a variety of materials to store heat. There is the development of the first generation of batteries for electric cars. Where such technological developments were once focused on producing smaller and smaller batteries for mobile phones and laptop computers, the enthusiasm of the major car companies for electric cars has given the research a new urgency. It is an obvious area in which to concentrate research funding.

Electrification of transport

This brings us to the impact of electrifying transport, and hence the opportunity to begin to get out of oil. There are several dimensions to this transformation. First, there is the technology – and the batteries. It is not a question of whether electric cars are feasible, but rather over what times-cale and at what cost. This is likely to be a continuous process of improvements.

The costs of electric cars have a major system component, and one that is poorly understood. The benefits of this technology come not only in the replacement of oil, but also in terms of the impact on the electricity system. Electric cars could comprise a massive electricity storage system. Car batteries store electricity in the way that cars currently store fuel. Virtually all cars carry a store of petrol or diesel. Very few are empty. Add this up over the vehicle fleet and it comes to a big number. Now think what this would mean for electricity. It would not just be the electricity itself that is stored in the batteries, but also the method (and timing) of the battery charging which could be transformational. Batteries can be charged at times when other demands on the system are low and when the supply is high.

In principle, this car charging could be counter to system demand, and hence its marginal costs could be very low. Indeed, at points of very tight margins on the system, the batteries could even release electricity back onto the system. Cars could therefore help to balance the system.

To this obvious enormous advantage there is a further benefit, which is the fit with intermittent renewables. We noted that when the wind blows, it pushes lots of electricity onto the system, but when it does not, back-up generation is required. Car batteries could be charged when the wind speed is good, and run down when it is not. This would not offset all the intermittency, but it would make an impact at the margin. But with car battery storage, direct electricity storage, and system demand management, intermittency may at some future date cease to be such a significant problem. Smart information technology would be complementary.

How would this work in practice? On the dashboard, the car's computer could report the charge in the battery and the remaining range for the vehicle. It could provide a number of charging options, in real time, taking account of the electricity generation on the system. It could also forecast future electricity generation, integrating weather forecasts to predict wind turbine output. You might then have three options: immediate charging; night-time charging; and options for intermediate solutions – all with real-time prices attached. The driver might be going to the supermarket, and its car park might have electric charging points. If there happened to be high wind generation that day, the car could be charged whilst you shopped, without compromising the other demands on the system, and therefore commanding only low prices. Eventually the choice may be made for you automatically. The way it would actually work depends on the infrastructure for battery charging and how it is managed. But the point remains: electric cars are radical in their impacts, with pay-offs not just in carbon but also in the wider system management – and therefore again in carbon.

Electric cars are not the only way forward for transport. They compete with fuel cells and hydrogen, as well as with gas-fuelled cars. Some argue that fuel cells rather than battery-powered cars will take off. Yet both have huge implications for the electricity industry. In an important sense, fuel cells are really just another form of battery, but the system implications are very different. It is too early to tell which will dominate.

New electricity generation technologies

The holy grail of decarbonization is finding new ways of generating electricity beyond conventional wind and solar. The laboratories of universities and major companies are full of people trying to do just that on a scale that is a step jump from what has been going on for decades. Bright engineering and science students are piling into alternative energy sources, and the results are confusingly encouraging. My email inbox is full of excited reports of the latest 'breakthrough'.

As with many aspects of climate change mitigation, it is impossible to pick the winners in this technology race. There will of course be winners – and losers – but it is in the nature of R&D that we do not know what the outcome will be. The best we can do is identify classes of technologies that look like good prospects. Amongst just some of the possibilities are next-generation solar, biotechnologies and possibly nuclear.

Solar has one obvious advantage: the sun comes up every day. Our planet has been driven by photosynthesis. This natural process has long been studied, and it is conceivable that it could be reproduced artificially. Indeed, artificial photosynthesis is gaining research momentum.[4] Solar represents the opportunity to tap directly into the earth's main energy supply, and in principle this leads to the possibility that, for all practical purposes, the energy supply is infinite. Unless the sun fails to come up one day, the energy potential is there to harvest. We are not going to run out of energy, and in that sense saving energy, and thereby reducing demand, is not really necessary. The problem is how to capture it.

Biotechnologies are really indirect variants on the solar theme. Biomass is produced by photosynthesis, which could be thought of as a CCS technology in reverse. It captures carbon from the atmosphere. Burning it releases it back, but at a much faster rate than would otherwise be the case. The trick here is not so much the return of the carbon to the atmosphere, but the ability to use biological processes to store the carbon – the very process that created our oil, gas and coal deposits. In terms of getting photosynthesis to provide energy via burning, there are a variety of technological options. As noted in Chapter 4, land and water are key constraints on current biomass options. Algae are one possible source that can overcome these constraints,[5] but there are a host of plants being considered in plant science laboratories, as well as parts of plants that we cannot currently use.

On the nuclear front, there has been no shortage of money for research. The problem is the sheer scale of the projects and their costs. There are ambitious projects relating to the concept of fast breeder reactors. Some of these ideas promise limitless cheap energy. But they have done this repeatedly in the past.

Finally, there is geothermal heat. It is already being exploited in volcanic locations like Iceland and New Zealand. The challenge is to tap geothermal heat more generally, and that requires a host of technological developments. As with solar, it could even be a practically unlimited source of energy supply.

This brief sample of ideas tells us two things. First, we should be technological optimists: climate change is not an insuperable problem, and there is no shortage of potential energy. Second, the world may look very different in a couple of decades. Energy policy has a quite natural tendency to view the future as if it is an extension of the past, and to project forward that past in terms of the known technologies. Fortunately that assumption is a poor one. No doubt, were our historian writing this chapter in 2050, he or she would be focusing on technologies that have been partially or completely ignored here. There will be surprises, but it is not enough to simply stand by and hope they turn up.

The problem is getting from here to there, and that requires money and effort to support R&D. Resources are scarce, particularly in economically difficult times. There are choices: money spent on offshore wind today is money not spent on R&D. And there are trade-offs to be made. In allocating resources to R&D, it is important to design R&D policy such that, on the one hand, there is no attempt to overtly pick winners, but that, on the other, resources are concentrated in promising areas. It is a delicate balancing act – identifying rich seams without picking winners.

R&D policies and the problem of market failures

Technological progress is not neat and precise. It involves bright people and sometimes eccentric entrepreneurs trying out ideas, alongside the more conventional larger firms and universities, with their laboratory practices of routine and order. Breakthroughs are sometimes the result of sudden shifts, and sometimes just the slow grind of trying to evolve towards more efficient and effective products.

There is therefore no obvious and correct method of moving from a decision to make progress and actually making it happen. If there were, it would all be straightforward. The large company model has sometimes delivered but, even in the pharmaceutical industry, the role of small innovative companies has often been decisive. In the shale gas case, it was the ability to put together three separate technologies – horizontal drilling, seismic information technology and fracking – which accounted for the breakthrough. Often this was driven by small entrepreneurial oil and gas companies, rather than the majors.

There is little that governments can do to create the sorts of people who will deliver these sorts of outcomes. Business schools run entrepreneurial classes, but there is little evidence that they produce better entrepreneurs. But what governments can too often do is get in the way, obstructing innovation. On the constructive side, they can help to coordinate and build supply chains. They can support universities and laboratories, and they can offer prizes – the classic example being Harrison's clock, the prize awarded by the Board of the Admiralty in the eighteenth century for a solution to the longitude problem in navigation.[6] And, of course, they control much of the research funding.

The usual way to think about R&D policy is to focus on market failures, and to seek to overcome them. These include the fact that R&D involves special kinds of risks, and lots of failures. These upfront costs are 'sunk': they typically cannot be recovered if the project fails. Most are wasted. Governments can help with direct funding, imposing levies on customers' bills, and granting generous tax allowances to write off the waste.

More importantly, knowledge is what economists call a public good: once it has been discovered, it costs nothing to provide it to everyone. The result is that without intervention, copying is rife and undermines the incentives to make the discovery. Here patents and the protection of intellectual property rights are crucial for the private sector. The alternative is the public provision of R&D funding through grants and through supporting and financing research institutions.

A more recent example is the competition to win a subsidy to develop CCS technology. This approach has been used in a number of countries. In Britain it has been a fiasco: a prize was offered, the competition reduced to one project, and then the sole remaining bidder pulled out. This was due in part to the prize being insufficient to underwrite the investment in the CCS demonstration project. But it was also due to confusion about whether the aim was to encourage private competitive advantage – to steal a march in the technology race (including possible patents) – or to provide the results as a public good to the world, and hence assist in the diffusion of CCS technology across developing and developed countries alike. The competition was relaunched in 2012, but still with little clarity of its core rationale and purpose.[7]

This example demonstrates the need to take account of the fact that R&D interventions are likely to display government failure as well as market failure. Promoters of technologies will see a potential R&D pork barrel, and they will use lobbying to try to capture the economic rents.[8] Given the complexity of sorting out the potential winners, it is not surprising that government failures have been rife.

These government failures matter because of the urgent need of R&D to deal with climate change, and because there are so many competing proposals for how governments should try to speed up the process. Amongst these proposals are, at one extreme, a new, Manhattan-style project at the global level,[9] and at the other, micro-subsidies for specific technologies. In between are a host of options, from a 'European MIT' (Massachusetts Institute of Technology) through to the introduction of a scale of subsidies for renewables support, determined according to the weight given by governments to specific technologies.

Where will the money come from?

Putting the policy design issues to one side for a moment, whatever approach is taken, R&D is expensive and the scale of the decarbonization challenge dictates that it will need a lot of money. The pay-offs are over decades, and so there is also urgency in the race against rising emissions.

Where will this money come from? If it is government, then total budgets tend to be fixed, and it will have important zero-sum-game elements: money for R&D means less money for elsewhere. The share of government expenditure going to R&D in developed countries is small, and in a number of cases biased towards military concerns. The share is higher in some developing countries, but their economies are smaller (although the gap is closing for some).

The budget constraint encourages government to raise money from the consumers via levies and taxes. This disguises the costs but arguably assigns them to the ultimate beneficiaries. Yet consumers, like governments, have limited resources: they have to be able to pay, and they have to be willing to vote for politicians who will force them to pay. The income effect of the levies and taxes matters.

These constraints have a very important policy implication: there is a limited amount of money available to spend on climate change mitigation in aggregate, whether it is government or consumers providing the subsidies. Whilst politicians like to promise more wind, biomass, and solar and lots of R&D too, the reality is that there are trade-offs. Money spent subsidizing existing deployable renewables is money that could have been spent on R&D. The policy approach therefore needs to start with the maximum amounts of money that consumers are willing and able to pay, and then to work out how best to spend it.

The good news is that there is every reason to believe that these major technology trends – communications, batteries, electrification of transport and new generation technologies – will probably transform both the demand and supply sides of the energy markets, and provide new low-carbon options. There probably will be others too. Together they may render the electricity and transport sectors as unrecognizable from today's experience as the communications sector is today compared with the early 1980s.

Rapid though some of these developments already are, there is a question over whether they will be developed in time. This is where governments come in, at both the national and international levels. It is where almost all of the climate change efforts should be concentrated, and fast. There is no time to waste on pet projects or on pandering to vested interests. R&D creates public goods, and these need to be distributed as rapidly as possible around the globe, given that climate change is a global public bad. Rather than devote so many scarce resources to things like offshore wind, it would be better to divert funds to R&D, and much better to do this on an international basis.

Conclusion

The naturalist Edward O. Wilson concluded his collection of essays *In Search of Nature* with a provocative chapter, 'Is Humanity Suicidal?':

> Unlike any other creature that lived before, we have become a geophysical force, swiftly changing the atmosphere and climate as well as the composition of the world's fauna and flora . . . No other single species in evolutionary history has ever remotely approached the sheer mass in photoplasm generated by humanity.[1]

It takes a naturalist to appreciate the enormity of the experiment we are currently conducting on our planet. Given the prospect of 9 billion people, a wall of consumption, and all the pollution and land use that goes with it, it is not hard to be fatalistic. Yet Wilson is an optimist, despite having spent his career watching and documenting the destruction of nature. But not if we go on like this. This is one of his main conclusions, and it is mine too. It is not beyond the wit – and the sheer ingeniousness – of humans to crack climate change.

Our current path is not getting us anywhere. The emissions keep going up, and nothing of substance has yet been achieved after more than two decades of effort. Faced with such a global threat, world leaders have been found wanting. All the international summits and agreements have been little more than showmanship. Leaders wring their hands in the global media and claim they are taking action. Bill Clinton, Tony Blair, José Manuel Barroso, Angela Merkel – they all claimed to be getting on with mitigation, but they have achieved almost nothing. They have been good at setting aspirations and targets, but just can't seem to deliver meaningful results.

None of this should be at all surprising. For it is based on what to all intents and purposes is analogous to a fairy tale. Along comes what could be one of the greatest threats in human history, and it is claimed that the energy sector can be converted from an overwhelmingly carbon-based one to almost zero carbon in the space of half a century, all at just a small cost, or even a profit, based on simplistic assumptions about economic growth and a 'new industrial revolution'. For some in this fairy-tale world, we do not even need new technologies, but can achieve it all on existing technologies – namely current renewables and energy efficiency. The green NGOs and green parties argue that nuclear is unnecessary. Energy efficiency will, we are told, reduce demand to such an extent that it will absorb any additional costs from the (temporarily expensive) renewables. The fantasy is that it will cost us little or indeed nothing more than we would have paid anyway, and this is a deceit that lies at the heart of the failure to achieve much over these two decades of trying.

It is easy to see how seductive politicians find this apparent win–win scenario. It is stitched together by a conventional wisdom of ever-rising fossil fuel prices and the spectre of economic chaos as oil runs out, so that however high the costs of current renewables might be, the future would be worse without them. If only it were that simple.

It isn't, and the cracks are there for anyone prepared to look. Whilst our leaders trot out the same old mantras, a potential disaster is unfolding around us. Anyone who listened to the political spin after the Durban climate conference can see the yawning gap between the hype and the reality. It is worth reiterating what actually happened: the world's leaders agreed that they would try to agree by 2015 what might happen after 2020.

A low point was when the then British Secretary of State, Chris Huhne, came back to the House of Commons and told MPs that 'we have achieved great progress in getting real action', and that 'we got every one of our asks'. For this, his fellow MPs on all sides congratulated him. It also bears repeating that you just couldn't make it up.

We remain committed in Europe to spending large amounts on a crash course of current renewables such as wind and some solar, and to a Kyoto framework based on carbon production targets. A Europe deep in the mire of an economic crisis, deindustrializing in any event, has piled on the costs at a time when its main international competitor, the US, has been reaping the economic windfall of historically low gas prices – in a context where the problem is not too few, but rather too many, fossil fuel reserves, of which the shale gas revolution is just one part. Anyone deluding themselves that we will be forced to decarbonize because we won't have enough fossil fuels needs to looks at the facts. That isn't going to happen in the relevant timescale for tackling global warming – if ever.

Europe's carbon consumption continues to increase, masked by the carbon production numbers. Remarkably, even in a recession like this one, Europe's carbon production in the last year has actually gone up too. Lots of money has already been spent; much more spending has been committed to the cause; and all with no apparent impact on global emissions. To achieve one of these outcomes would be a serious mistake: to achieve both simultaneously is indeed a disaster. Worse, to do so in a way that renders our energy supplies less secure and undermines our competitiveness merely adds to the absurd situation we have got ourselves into. In the hall of political mistakes, this must stand out as a pretty unique failure.

This political narrative takes as its starting point the assumption that climate change will get solved in Europe – at least initially. Europe will show the way so that others may follow the lead. As a result, almost all the efforts are going into trying to reduce carbon production in Europe, and none are going into reducing carbon consumption. Few politicians stand up and spell out that climate change is all about coal, economic growth and population growth. It is as if the three new coal power stations being opened every week in China and India don't matter, as long as we open a few wind farms in Europe and install some insulation in our houses. It would be unkind to call it a 'windmills-and-draught-excluder strategy', but there is a painful grain of truth in this characterization.

It is not as if we have time to muck about. The clock is ticking. Urgent action is needed: since 1990 we have been adding carbon to the atmosphere at an accelerating rate, rising from 2 ppm per annum to now 3 ppm. Overall levels will soon hit the 400 ppm mark. Hoping it will all go away is not going to work, and nor is pretending that mitigation will be cheap. Contrary to what many politicians would wish us to believe, energy policy cannot deliver both sustainable and cheap energy simultaneously, regardless of whether it is secure. When Tony Blair wrote in 2003 that Britain's energy policy was 'cheap energy, that is also sustainable and secure', he was deluding not only himself but the wider public too.[2] He was not alone: all across Europe we have been led to believe in this miracle, but it is just not happening.

The challenge for the current crop of world leaders is to face up to this failure, and to square with their voters. The cost of the decarbonization of entire economies is likely to be very high, and it is going to involve sacrifices. That is the real message that comes from seeing carbon consumption, rather than carbon production, as at the root of the problem – including all our spending on those energy-intensive products we once made in Europe and now import from countries like China. These are our emissions too, and if we were to really decarbonize our consumption, nobody would be talking about the almost free lunch our leaders promise us. It's a lot tougher to be really green in a climate change sense – and indeed tougher too if we add in all the rest of the environmental destruction that comes from our high-consuming ways. When villages, towns and even cities claim to be moving to being 'zero-carbon communities', it is worth taking a walk around and seeing what is really going on. Are people buying only zero-carbon-produced goods in the shops? Are they eating only zero-carbon food (whatever that means)? They may have local, apparently low-carbon, energy sources, but that does not mean that they have zero-carbon consumption.

Consider what is being proposed. The current industrial structures of all developed and most developing economies are overwhelmingly carbon-based. Power stations, oil refineries, gas and electricity networks, cars and road transport, industrial plants, and aviation – these all embed carbon in the economy. Decarbonizing requires the coordinated replacement of almost all of the capital stock – of the world.

The nearest analogy is the conversion of peacetime economies in the 1930s to wartime economies by 1940, but even this fails to do justice to the scale of the transformation required. Switching from making cars to making tanks is an evolutionary step, whilst switching from petrol-fuelled cars to electric ones needs a combination of new battery manufacture, a national network of charging stations, and a reconfiguration of the electricity system to cope with the scale and nature of the new demands. And that is just cars. The same sorts of transformations are needed for all the other major industries too.

It is true that this new capital stock will create jobs to make the various capital investments – just as tank-making and other military demands put people back to work in the late 1930s. But militarization was largely a home-made business: the new green jobs are as likely to be in China as in Europe. It is also the case that, over time, the capital stock needs to be replaced anyway, so some of the transformation is not simply scrappage.

Almost every developed country claims that it can and will be a leader in the renewables business. In his 2012 State of the Union address to Congress (24 January), President Obama claimed that the US 'will not cede the wind or solar or battery industry to China or Germany'. But all cannot be winners, and much of the design, manufacture and assembly work is rather similar to many other manufacturing processes. Europe has been deindustrializing. It has been doing this for a very good reason: it cannot compete in mainstream manufacturing with developing countries like China. Why suddenly the basic economics that have been driving this restructuring for the last couple of decades should not apply in the particular cases of wind and solar is hard to fathom, especially given that China is in fact already out-competing European wind and solar manufacturers

– so much so as to trigger widespread failures and even bankruptcies for some US and European solar companies.

Regardless of whether all these green jobs materialize, there is a darker side to Europe's obsession with current renewables: the so-called brown jobs (jobs in the energy-using industries) are destroyed in the process. It may well be that Europeans have already given up on many energy-intensive industries, preferring to import rather than produce. But the substitution still continues, dampening the overall job numbers, as these industries are out-competed by America, China and India, and businesses choose to locate away from the high costs of Europe. Politicians are always keen to point to green investments creating a given number of new jobs – but they almost always quote gross, not net, numbers. In our example of switching from peacetime to wartime production, it was the case that this led to full employment. But it is hardly a state of affluence. Neither would a wind farm state be affluent. Tanks and windmills have quite a lot in common.

Confronted with a war, Winston Churchill informed British civilians that he offered nothing but 'blood, toil, tears and sweat'.[3] The public understood this, and responded by accepting a sharp curtailment of their consumption. Climate mitigation does not require blood, sweat or tears (although there may be some toil involved), but it will have an impact on consumption, so that savings can be made available for the huge investment effort required. This is the real inconvenient truth – the carbon crunch.

Telling the truth is an important starting point. A reality check has in any event materialized as the costs of the renewables in Europe start to show up on consumers' bills. Politicians have suddenly begun to focus on how to address the bills issues. Bashing the energy companies (like bashing bankers), and blaming high prices on their alleged greed and anti-competitive behaviour, is a natural political reflex, but it lacks credibility. In Britain, the current Secretary of State, Ed Davey, puts the bills issue down to world gas prices. In the short run he may be right, but that is not the point. Everyone can see that there are price increases coming to pay for all the subsidies that have been doled out on their behalf to wind farm developers, land owners, and those quick enough to get in on the super-subsidies for rooftop solar – not to mention the numerous levies that pay for insulation. This is just the beginning of a backlash that was foreseen by some – including the author – as the promise of 'no pain' was revealed as a sham.

This backlash is dangerous and it will have serious consequences. If the politicians have misled us about the bills, then it is an easy step to doubt what they say about the science behind climate change. It encourages climate scepticism to fester. Dire warnings of Armageddon have a horrible habit, like Jehovah's Witnesses' predictions of dates for the end of the world, of being found out. Treating every major weather event – provided that it is on the hot or violent side – as evidence of climate change, but keeping quiet when there are cold winters, does little to add to credibility. Trashing those who do not 'believe' in current renewables policy, or who question the scale of energy efficiency opportunities, merely reinforces the suspicion that climate change is more religion than well-supported hypothesis.

The green NGOs and the green parties have a particular responsibility for what has happened. They have had few scruples about exaggerating to make their points. Their websites are full of hyped-up scare stories, and they often play fast and loose with the science. But as their political wings have grown, so too has the reluctance to spell out the full implications of their policies. Returning to what many of them regard as a small-scale sustainable set of communities, free of nuclear power and free from the hated incentives and profits of capitalism and 'big business' (whatever that is), they suggest that we can feed, house and supply energy to 9 billion people on a diet of wind farms, solar, energy efficiency and related technologies, and that in Europe we can do this in the space of a couple more decades.

Many would argue that this isn't credible, but that is to miss the point. It's an ideology, and it has its own coherence. It would be a very egalitarian world, with equal carbon rights to the fore. In carbon terms, it would treat everyone equally, regardless of where they lived, and regardless of when they lived – a view which is shared with Nicholas Stern, with his zero time-preference rate. It is a deeply redistributionalist view, and it is way beyond the sorts of adjustments mainstream socialists call for. We are not talking here about changing the top rate of tax by a few percentage points: we are talking about redrawing the income map of the world on a drastic scale. And of course this would not be the outcome of some democratic choice: it would have to be imposed. It is extraordinarily radical, and arguably goes against the grain of human nature. It has not the slightest chance of being effected in the time period within which climate change needs to be cracked – if ever. In the meantime, it mandates the forcing-through of technologies that fit with the ideology. Greens are very good at telling us what they are against. The trouble is that there is not much left. They are against nuclear of course, they don't like coal and oil, and they really hate shale gas, since they rightly realize that it is a direct threat to the economics of their preferred wind farms.

Scientists have not always helped the cause of tackling climate change. A wary public witnesses what the media scientists say (it is, after all, mainly how they get to see what scientists do) – the threats of Armageddon, the insinuations that extreme weather events are evidence of climate change, and of course the occasional glimpses of the way they go about peer review. What the public sees is all too human, and often rather different from the idealized picture of the disinterested, public-spirited scientist. Economists have added to this credibility gap by claiming that the costs of mitigation may be low and that they may not be detrimental to living standards. The Stern Review is a clear case in point.

The conclusion that we are in a hole of our own making is hard to escape. But it need not be like this – we can stop digging. Starting from an unrelenting realism about the causes of climate change – and the costs of tackling it – and recognizing that consumers and taxpayers have limited resources, putting a very sharp focus on the least-cost opportunities for mitigation offers a very different and much more positive path.

Setting the fairy tale aside, what is needed is a realistic energy policy based on a recognition that resources are scarce, and that spending on one technology or particular support mechanism means that this spending cannot be targeted at another. This simple reality passes many politicians and advocates of current policies by: they in effect assume that there are infinite resources available, and that we can pursue all the options simultaneously. A flavour of this was given in President Obama's 2012 State of the Union Address in the context of energy security. He promised 'a future where we're in control of our own energy', and he stated that 'this country needs an all-out, all-of-the-above strategy that develops every available source of American energy.' It represents a special kind of economic illiteracy, and a political side-stepping of the difficult choices that have to be made.

How much have we got to spend on climate change? This is the sort of question environmentalists abhor: how could mitigating such a threat be limited by money? Yet it is, and the way to think about the problem is to look at the basis for the investment programme that might be needed. If the whole capital structure has to be replaced, and the transport sector has to be electrified, we can work out very roughly how much this might cost and then apply this to the population. Since it is investment, it will need to be met by savings – either now or in the future (we return to the question of debt finance below). Savings are what are left from income after consumption. So more savings (and more investment) means less consumption.

So far the logic is pretty straightforward. But now comes the political and economic consequence – the carbon crunch. Less consumption means lower standards of living. It is no longer the fairy-tale world set out above: *decarbonization cannot be done with zero pain*. On the contrary, the sorts of investment programmes required could be very painful indeed. That is precisely where the war analogy is most telling: in moving from a peacetime to a wartime economy, room in the economy had to be found for the military production, and this was partly effected by squeezing consumption. Indeed it was so squeezed that rationing was widespread.

Tackling climate change does mean lowering our standard of living from its current unsustainable levels, even after the economic crisis. There is no escaping this fact. But once we recognize that current consumption levels are not sustainable, it places an enormous premium on finding the most cost-effective ways of reducing emissions, and taking account of the time periods. But before we get to this, there is one counter-argument that needs to be addressed: the idea that we can solve the problem of financing all this investment through borrowing. It is a fashionable approach: it is argued that consumers can borrow to make energy efficiency investments; companies can gear up their balance sheets with debt to finance new power generation; utilities can borrow to finance new infrastructures; and that governments can borrow to support any of these investments that the private sector cannot.

The problem with debt, as the developed world, and particularly the Europeans have discovered, is that it has to be repaid. And there are only two parties who can repay it (and the interest that accrues): consumers and taxpayers. At the aggregate level, consumers and taxpayers are, for practical purposes, the same. Borrowing postpones payment: it does not remove it. There is no magic climate change free lunch.

This sort of political deception has probably run its course. The credit crunch put paid to the idea of easy borrowing, and for most developed countries it is now consumer levies that provide the revenues to back the energy-related debt. And there is no evidence to date that consumers are prepared – or able – to pay very much.

How then do we persuade people that they must cut back on their consumption, and do so in ways sufficient to fund all the investments required to tackle climate change? We cannot avoid the costs ultimately falling on the public – as taxpayers or consumers. That cost is the (global) damage their carbon consumption causes. Economists tend to be very keen on introducing a carbon price to reflect this damage. The case is extremely powerful. The price influences the choices every household, business and government makes. It has a substitution effect (it incentivizes us to switch away from carbon-intensive goods) and an income effect (it reduces our ability to consume). It allows the market to find the cheapest ways of reducing emissions, free from all the lobbying and vested interests; and it bears down on governments when they are foolish enough to try to pick winners. There is no hiding from a price.

What precisely the tax is levied on matters a great deal. Carbon consumption is what matters, and explicitly or implicitly raising prices and costs at home with a carbon production tax, but not for imports, leads to carbon leakage. A border carbon adjustment is essential if the tax is to do its work properly.

Europe has opted for permits instead: the EU ETS. It is seriously flawed, being short-term, volatile and yielding a price so low as to have little impact. It could be ramped up, and its politically seductive characteristics may well lure other countries down the emissions-trading route rather than opting for a tax. But it would be better to go for a longer term, stable and rising carbon tax. To bridge the gap, a price floor would probably do the job, taking the strain gradually, so that the hapless EU ETS might slowly wither away, saving the blushes and all the political capital the politicians have invested in it. The point of using border adjustments is that with scarce resources to devote to mitigation, we don't want to waste them – which is what happens with carbon production-only targets and the associated instruments.

A border carbon tax has one other great virtue. It provides a bottom-up way of getting global action, and avoids the tortuous Kyoto-style top-down negotiations. Much is made of a 'coalition of the willing' – which in practice means the Europeans – and we can either make this work by imposing carbon pricing domestically and at the border, or recognize that good intentions are largely futile. Border carbon adjustments have the useful property of encouraging others to implement their own carbon-pricing measures in order to keep the revenues that would otherwise go to the importing country. In principle, it can be done in quantities or prices – by emissions permits at the border or by taxes. In addition, standards can be applied to imported goods and services that apply domestically. This is already widely undertaken in the case of car pollution and efficiency standards.

Applying border adjustments is bound to take time and cause transitional difficulties. Carbon pricing is necessary, but it is not sufficient. In the meantime, we need to again think hard about not wasting resources. It is here that the 'no regrets' concept comes into play. Are there things we could do that would be a good idea anyway, even if there were no climate change to worry about? And nobody else acted? The good news is that the answer is a strong yes. There are very good reasons for replacing coal with gas, regardless of climate considerations. Gas is cheap and abundant. It increases security of supply in many cases, improves competitiveness, and does not increase consumer bills. It is less capital-intensive than many current renewables technologies and nuclear, both of which tend to be overwhelmingly capital-intensive. Finally, gas can be both baseload and flexible, so if it does turn out that we need lots of intermittent generation then it will be an advantage to have gas on the system. Even better, governments do not have to do much to make this happen, other than ensuring that the electricity market does not penalize gas investments, and avoiding putting too many regulatory hurdles in the way.

The US has begun to show the way. Long the target for criticism by the green NGOs, and indeed European politicians generally, when it comes to climate change, the US did not ratify Kyoto. They have not accepted a top-down carbon production target. You might think that, as a result, their carbon record would be much worse than that of the Europeans. But it isn't. From a production point of view it is falling fast as it takes up the opportunity to switch from coal to gas for power generation, whilst Europe's carbon production is no longer falling, despite the economic crisis. Over the coming decade it is entirely possible that the US will make significantly greater inroads into its carbon production than the Europeans. There is one further difference: the US trades a lot less than the Europeans, and therefore the impact of imports on its carbon consumption is correspondingly smaller than for many European countries.

Compare the impact of this strategy with the case for going flat out for current wind and solar technologies now. These are, by contrast, likely to have significant impacts on consumer bills and require significant government subsidies (including exemptions from taxes, tax-preferential treatment, and a host of related support mechanisms). Indeed, so great are the costs likely to be in the short term, it will probably leave few if any consumer resources for other aspects of climate change mitigation. It will exhaust already stretched energy budgets. We will be putting not only all our eggs in the single wind and solar basket, but all our scarce money too. There will be precious little left for anything else – including adaptation.

This matters because there is a further inconvenient fact that has to be faced. Even if we devoted all our resources to current wind and solar technologies, they would not be anything like enough to solve the problem of climate change. There simply is not enough land and shallow seabed, given how little intermittent power each wind turbine creates. Even desert solar with long-distance transmission links would not be enough. An inescapable conclusion follows: we will need new technologies.

New technologies do not grow on trees: they take lots of resources to discover and develop, especially on the large scale needed to address the rising energy demands that come from economic and population growth. Whilst we should take transitionary measures where these are cheap and 'no regrets' now, we should also recognize that money spent on large-scale wind will be money not spent on R&D.

R&D is also very much a wasteful business – lots of what look like good ideas turn out to be 'dogs'. Whilst on the one hand this requires a promiscuous approach to trying out ideas and concepts, it also means that there are huge gains in pooling R&D resources. Think what could be done if every country in the world devoted, say, 0.5% of their GDP to a global climate mitigation R&D fund, let alone Stern's 1% per annum. Even 0.1% aggregated up would be a huge sum.

The resources saved by concentrating on gas rather than current wind and solar could be put to better use by focusing on the longer-term climate change challenge. Existing technologies cannot solve the problem, but again there is good news: the potential of new technologies is awe-inspiring. From the near-term options like batteries, information technology applications and electric cars, to the potential breakthroughs in technologies like artificial photosynthesis, the research world is alive to the energy challenges. On the R&D front, the scope for surprises, almost regardless of what policy-makers do, is considerable. Much of the energy research is unstoppable, and in any event commercial companies will look for opportunities. But what is needed is money – lots of it – and this is best achieved by pooling resources internationally.

By considering climate change mitigation in the context of a limited ability to pay – and a limited willingness to pay – a very different strategy emerges. For the amounts currently spent, carbon emissions could be cut by much more in the short run, and the possibility of a series of breakthroughs could be husbanded through a major international R&D programme. In the transition, gas-for-coal is largely no-regret, and it could be argued that a major international effort (or at least a coalition of the willing) would be largely no-regret too. Add in the carbon price with border adjustments, and the stage is set for making progress rather than ineffectually spending money whilst passively standing by as global emissions continue to grow. It will nevertheless still be painful for our standards of living – but there is no getting away from this. This pragmatic approach would be better (probably much better) than what is going on now, but even if it were implemented quickly, we are in for quite a bit of climate change anyway. Warming by at least 2°C seems almost inevitable. Therefore, there will have to be adaptation, and whilst it is understandable that many green NGOs blanch at the mere suggestion of adaptation for fear of abandoning mitigation, it will have to be faced.

These three elements together represent a plan – but one that involves eating a sizeable chunk of humble pie. It involves an admission not only of failure, but also of misleading the public about the costs of the policies that have been adopted. The next generation will inherit a warmer world, and they will not thank the politicians who claimed that the costs would be low, and that they were showing world leadership. Furthermore, they will not necessarily be impressed by the sorts of analyses that supported these claims.

What are the chances of this alternative approach being taken? At first glance they might seem pretty low. But there are some glimmers of hope. The gas will come anyway because it is least-regret. The US in particular may reduce emissions much faster than Europe over the coming decade as it switches from coal and exploits its abundant shale gas resources – even on a business-as-usual basis. It will do this for economic reasons and in pursuit of energy security and independence. Lower international gas prices would help this transition elsewhere, and China will go for shale gas anyway.

These are very real grounds for optimism. There will in any case be less support for the current renewables and solar efforts, and even here the folly of putting all the eggs in these particular 'winning' baskets is already becoming apparent. What will further constrain these political projects is the sharp collision between the policies and consumers. Politicians need to be reminded of the two golden rules of energy policy: consumers must be able to pay; and if they can, they must also be willing to vote for politicians who will force them to pay. The job of politicians is to explain to their electorates what those costs are and why they should pay them, and to have a credible plan that addresses the problem in ways that make the most of their money. This is the political face of the carbon crunch.

It will be for future historians to reflect on whether humans rose to the climate change challenge, or whether they are indeed suicidal. So far, there is little to show for more than two decades of effort and cost. Kyoto has achieved nothing of substance, and the chance to limit emissions to 400 ppm has effectively been squandered. World leaders have a lot to answer for, and it is doubtful that history will judge them kindly. There remains hope that at this late stage, effective action will be taken. Climate change is a problem that can be cracked – but it won't be on current policies.

Endnotes

Introduction

1. For ease of style I refer throughout to carbon, rather than carbon dioxide (CO_2). Where I refer to ppm, this means CO_2, and not carbon dioxide equivalence, CO_2e. CO_2e incorporates all the other greenhouse gases. Where this is relevant – in quoting the Stern Review, for example – it is explicitly stated.
2. The 2°C figure is usually associated with 450 ppm CO_2e – it is less if measured in CO_2, as in the main text here.
3. Hedegaard, C., 'The EU's Role in Fighting Climate Change', in Barysch, K. (ed), *Green, Safe, Cheap: Where Next for EU Energy Policy?*, Centre for Economic Reform, London, 2011.
4. Ibid., p. 39.
5. Ibid., p. 39.

Chapter 1: How serious is climate change?

1. For example, Roland Emmerich's *The Day After Tomorrow*, 2004, Al Gore's *An Inconvenient Truth*, 2006, and Yann Arthus-Bertrand's *Home*, 2009.
2. International Energy Agency, 'World Energy Outlook, 2011', November 2011, p. 40.
3. See www.spiegel.de/flash for a number of opinion polls on the position of the main political parties in Germany.
4. www.people-press.org provides a number of surveys on public opinion priorities in the US. See www.people-press.org/2012/03/19/as-gas-prices-pinch-support-for-oil-and-gas-production-grows/
5. See Yale Project on Climate Change Communication, 'Americans' Global Warming Beliefs and Attitudes in November 2011', and 'Public Support for Climate and Energy Policies in November 2011', Center for Climate Change Communication, George Mason University, November 2011. See also Weber, E. U. and Stern, P. C., 'Public Understanding of Climate Change in the United States', *American Psychologist*, 66(4), May–June 2011, pp. 315–328.
6. Lawson, N., *An Appeal to Reason: A Cool Look at Global Warming*, London: Duckworth, 2008.

7. Lomborg, B., *The Skeptical Environmentalist: Measuring the Real State of the World*, Cambridge: Cambridge University Press, 1998.
8. In particular from Phil Jones, head of the Climatic Research Unit (CRU); Keith Briffa, a CRU climatologist specializing in tree ring analysis; Tim Osborn, a climate modeller at CRU; and Mike Hulme, director of the Tyndall Centre for Climate Change Research.
9. An article in *Nature* appeared quickly after the European heatwave, which claimed that man-made emissions had doubled the risk of an event of this magnitude. Stott, P. A, Stone, D. A. and Allen, M. R., 'Human Contribution to the European Heatwave of 2003', *Nature*, 432(2), December 2004, pp. 610–614.
10. In the IPCC AR4 2007 Working Group Report it was claimed that the Himalayan glaciers could melt by 2035 or sooner. In 2010, after the Copenhagen COP, the IPCC retracted this claim. IPCC, 'Impacts, Adaptation and Vulnerability: Contribution of Working Group II to the Fourth Assessment Report of the Intergovernmental Panel on Climate Change', Parry, M. L., Canziani, O. F., Palutikof, J. P., van der Linden, P. J. and Hanson, C. E. (eds), Cambridge: Cambridge University Press, 2007.
11. IPCC, 'Renewable Energy Sources and Climate Change Mitigation: Special Report of the Intergovernmental Panel on Climate Change', Cambridge University Press, 2012.
12. Interview in the *European Energy Review*, 2 May 2012, available at www.european-energyreview.eu/site/pagina.php?id=3681. See also Vahrenholt, F., *Die Kalte Sonne* (The Cold Sun), Hoffmann & Campe Vlg Gmbh, February 2012.
13. Stern, N., *The Economics of Climate Change: The Stern Review*, HM Treasury, Cambridge University Press, January 2007.
14. Nordhaus, W., *A Question of Balance: Weighing the Options on Global Warming Policies*, New Haven: Yale University Press, 2008, p. 167.
15. Recall that CO_2e translates all greenhouse gases into carbon dioxide equivalence, and a very rough rule of thumb is to take off around 40 ppm from CO_2e to get CO_2. So 550 ppm CO_2e is around 500 ppm CO_2, and our current (roughly) 380 ppm CO_2 is around 420 ppm CO_2e.
16. Stern, *The Economics of Climate Change: The Stern Review*, 2007, p. 239. Note that Stern uses CO_2e, in which he incorporates the warming effects of other greenhouse gases, excluding the effects of aerosols – which make a great deal of (negative) difference.
17. The treatment of Bjørn Lomborg's book, *The Skeptical Environmentalist*, which did not in fact suggest that nothing should be done, but rather put climate change in the context of other global challenges, is striking. He faced ferocious academic and green hostility, and was frequently shouted down by those who had often not bothered to read his book. Nigel Lawson found it very difficult to find a publisher for his own sceptical book, *An Appeal to Reason* – the publishers feared a backlash. Paradoxically, the result was that both of these books sold well. Lomborg, B., *The Skeptical Environmentalist: Measuring the Real State of the World*, 1998, and Lawson, N., *An Appeal to Reason: A Cool Look at Global Warming*, 2008.
18. European Commission, 'Achieving a Deal on Climate Change: An EU View on Copenhagen Council of Foreign Relations New York', speech by José Manuel Durão Barroso, President of the European Commission, 21 September 2009. Available at http://europa.eu/rapid/pressReleasesAction.do?reference=SPEECH/09/401&type=HTML
19. Weart provides an excellent history of how scientists discovered global warming (Weart, S. R., *The Discovery of Global Warming*, London: Harvard University Press, 2003). Fourier had worked out that the earth's surface emitted infrared radiation, but it was left to Tyndall to discover why the earth was not left freezing as the sun's radiation was bounced back into space. Tyndall, having confirmed in 1859 that oxygen and nitrogen were transparent, then found that coal gas was opaque for heat rays, and went on to find that CO_2 was opaque as well. This was the greenhouse effect in a laboratory: CO_2 restricted the amount of heat that was dissipated into space.

20. Arrhenius worked out that less CO_2 would lead to a cooling effect. This cooling would result in less water vapour in the atmosphere. This in turn would lead to further cooling and an ice age could result. See again Weart, *The Discovery of Global Warming*, 2003.

21. Hardy, J. T., *Climate Change: Causes, Effects and Solutions*, Chichester: J Wiley, 2003, pp. 29–34 provides an accessible summary of the evidence sources.

22. This is a subject of intense debate, yet it is hardly surprising that forecasts of climate change applied to a time period as short as a decade are subject to considerable error. The issue is whether 'closed' science tries to reinterpret data to support the hypothesis, or is genuinely open to the alternative hypothesis about determinants of climate.

23. IPCC AR4 2007 Working Group Report. These have been distilled in four major IPCC assessment exercises, the latest in 2007, and underlying them are a host of scenarios, depending on the model characteristics (how they think climate is determined) and the assumptions about the background conditions.

24. The idea that science is best conceived of as a process of conjectures and refutations has been forcefully advanced by Popper, K., *The Logic of Scientific Discovery*, originally published in Vienna: Verlag von Julius Springer, 1935, and *Conjectures and Refutations: The Growth of Scientific Knowledge*, London: Routledge, 1963.

25. For recent evidence of the impact of climate change, see D'Andrea, W. J., Huang, Y., Fritz, F. C. and Anderson, N. J., *Abrupt Holocene Climate Change as an Important Factor for Human Migration in West Greenland*, Proceedings of the National Academy of Sciences of the United States of America, 2011, available at www.pnas.org/cgi/doi/10.1073/pnas.1101708108

26. The freezing of the Thames was a feature of the Little Ice Age, but it was exacerbated by a key weir at London Bridge, which limited the range of the salt water and slowed the flow. See Lamb, H., *Climate, History and the Modern World*, 2nd edn, London: Routledge, 1995.

27. The Little Ice Age may have been a localized effect as the systems of high and low pressure in the northern hemisphere changed places. One hypothesis is that four very large volcanic eruptions between 1273 and 1300 triggered a cooling process which became self-reinforcing. Miller, G. H., et al., 'Abrupt Onset of the Little Ice Age Triggered by Volcanism and Sustained by Sea-ice/Ocean Feedbacks', *Geophysical Research Letters*, 39, L02708, 5 PP., 2012.

28. There is a massive and much disputed literature on the costs of different levels of global warming. Richard Tol's work is a good source. See, for example, Tol, R., 'The Economic Effects of Climate Change', *Journal of Economic Perspectives*, 33(2), Spring 2009, pp. 29–51.

29. See Crawford, R. M., *Plants at the Margin: Ecological Limits and Climate Change*, Cambridge: Cambridge University Press, 2009.

30. In 2008 the US Geological Survey stated that 'These resources account for about 22 percent of the undiscovered, technically recoverable resources in the world. The Arctic accounts for about 13 percent of the undiscovered oil, 30 percent of the undiscovered natural gas, and 20 percent of the undiscovered natural gas liquids in the world. About 84 percent of the estimated resources are expected to occur offshore.' US Geological Survey, '90 Billion Barrels of Oil and 1,670 Trillion Cubic Feet of Natural Gas Assessed in the Arctic', news release, 23 July 2008. Available at www.usgs.gov/newsroom/article.asp?ID=1980&from=rss_home

31. See Fagan, B., *The Long Summer: How Climate Changed Civilization*, London: Granta, 2004.

32. Richard A. Posner's book *Catastrophe: Risk and Response* considers this alongside other events, such as an asteroid collision. Posner, R. A., *Catastrophe: Risk and Response*, New York: Oxford University Press, 2004.

33. See Wilson, E. O., *The Diversity of Life*, London: Penguin, 1994.

Chapter 2: Why are emissions rising?

1. Jevons, W. S., *The Coal Question: An Inquiry Concerning the Progress of the Nation, and the Probable Exhaustion of Our Coal-mines*, London: Dodo Press, 2008.
2. Ibid., p. 1.
3. Ibid., p. 98.
4. Ibid., p. 236.
5. See National Academy of Sciences, *The Hidden Costs of Energy*, Washington: National Academics Press, 2010. Greenstone, M. and Looney, A., 'Paying Too Much for Energy? The True Costs of Our Energy Choices', Working paper 12-05, MIT Department of Economics, February 2012, summarizes the evidence and aggregates the impacts.
6. See Ghose, M. K. and Majee, S. R., 'Air Pollution Caused by Opencast Mining and its Abatement Measures in India', *Journal of Environmental Management*, 63(2), 2001, October, pp. 193–202.
7. The Large Combustion Plant Directive (LCPD, 2001/80/EC) was introduced by the European Parliament and Council on 23 October 2001.
8. Chamon, M., Mauro, P. and Okawa, Y., 'Cars: Mass Ownership in the Emerging Market Giants', *Economic Policy*, 23(54), 2008, pp. 243–296.
9. IPCC, 'Aviation and the Global Atmosphere: A Special Report of IPCC Working Groups I and III', 1999.
10. International Energy Agency, *Coal: Medium-term Market Report: Market Trends and Projections to 2016*, December 2011, p. 9.
11. International Energy Agency, *Coal: Medium-term Market Report: Market Trends and Projections to 2016*, 2011, p. 31.
12. Nicholas Stern takes this plan seriously, and hence presents an optimistic picture of China's emissions profile. Stern, N., 'Raising Consumption, Maintaining Growth and Reducing Emissions: The Objectives and Challenges of China's Radical Change in Strategy and its Implications for the World Economy', *World Economics*, 12(4), 2011, pp. 13–34.
13. See International Energy Agency, *World Energy Outlook, 2011*, November 2011, Table 12.1, p. 451.
14. See Chapter 3 of Lane, N., *Life Ascending: The Ten Great Inventions of Evolution*, London: Profile Books, 2010.
15. US Energy Information Administration, *International Energy Statistics 2011*, Washington DC, 2012.
16. A feel for the scale of these new coal reserve developments can be gleaned from taking a look at the Galilee project in Australia. This is coal mining on a vast scale, combining a new rail transport system and a major port facility to export coal to China. See China First Coal Project Presentation, July 2011. Update, via www.waratahcoal.com
17. For a discussion of some of these threats and the problems associated with motivating global cooperation, see Barrett, S., *Why Cooperate? The Incentive to Supply Global Public Goods*, New York: Oxford University Press, 2007.

Chapter 3: Who is to blame?

1. Hardin, G., 'The Tragedy of the Commons', *Science*, 162, 1968, pp. 1243–1248, provided the classic statement.
2. See Collier, P., *The Bottom Billion: Why the Poorest Countries are Failing and What Can be Done About It*, New York: Oxford University Press, 2007. Chapter 7 discusses the complexities of aid as a way of promoting growth for the worst off, stressing both what is achieved, but also the importance of how it is delivered.
3. Rawls, J., *A Theory of Justice*, Cambridge, Massachusetts: Harvard University Press, 1971.
4. For an explanation of justice as impartiality, see Sen, A., 'Adam Smith and the Contemporary World', *Erasmus Journal for Philosophy and Economics*, 3(1), 2010, pp. 50–67.

5. For a discussion of the role of human nature as a consideration in discounting see Helm, D. R., 'Sustainable Consumption, Climate Change and Future Generations', *The Royal Institute of Philosophy Supplements*, 69(1), 2011, pp. 235–252.

6. Scruton, R., *Green Philosophy: How to Think Seriously About the Planet*, London: Atlantic Books, 2012.

7. Stern, N., *The Economics of Climate Change: The Stern Review*, 2007.

8. Ramsey, F., 'A Mathematical Theory of Saving', *Economic Journal*, 38(152), 1928, pp. 543–559. Although the Cambridge utilitarians were much influenced by G. E. Moore's idealism, the Ramsey approach to zero discounting of time was very much in the conventional utilitarian tradition. Stern displays similar utilitarian tendencies.

9. Stern, N., *The Economics of Climate Change: The Stern Review*, 2007, p. 35.

10. Ibid., p. 60.

11. See Wilson, E. O., *In Search of Nature*, London: Allen Lane, 1996. The last chapter is entitled 'Is Humanity Suicidal?'

12. Nozick, R., *Anarchy, State and Utopia*, Oxford: Blackwell, 1974.

13. Coase, R. H., 'The Problem of Social Cost', *Journal of Law and Economics*, 3, 1960, pp. 1–44.

14. Hillman, M., *How We Can Save the Planet*, London: Penguin Books, 2004, Chapter 7; and Monbiot, G., *Heat: How to Stop the Planet Burning*, London: Penguin Books, 2007.

15. On concepts of global justice, see Miller, R. W., *Globalizing Justice: The Ethics of Poverty and Power*, New York: Oxford University Press, 2010.

16. Hume, D., *Enquiries Concerning Human Understanding and Concerning the Principles of Morals*, reprinted from the 1777 edition with an introduction by L. A. Selby-Bigge, Third Edition, Oxford: Oxford University Press, 1975.

17. Blair is often portrayed as being a key 'leader' on climate change, yet he made remarkably little progress, including at the famous Gleneagles Summit in 2005. As can be seen from the very few references to climate change in his autobiography, he failed to persuade either President Hu of China or President Bush to do much. It is revealing that Princess Diana gets considerably more attention in the book. Blair, T., *A Journey*, London: Hutchinson, 2010.

18. These numbers are taken from Helm, D., Phillips, J. and Smale, R., 'Too Good to be True? The UK's Climate Change Record', 2008. Available at www.dieterhelm.co.uk

19. The corollary is that if carbon consumption is taken into account, world trade would contract. See Böhringer, C., Carbone, J. C. and Rutherford, T. F., 'Embodied Carbon Tariffs', National Bureau of Economic Research, working paper series, 17376, August 2007. www.nber.org/papers/w17376

Chapter 4: Current renewables technologies to the rescue?

1. For a thorough analysis of the costs of renewable electricity generation, see Borenstein, S., 'The Private and Public Economics of Renewable Electricity Generation', *Journal of Economic Perspectives*, American Economic Association, 26(1), 2012, pp. 67–92.

2. In mountainous areas such as Scandinavia and parts of the Alps, hydro can provide a complement to wind, and therefore make the economics rather different.

3. DUKES – the *Digest of UK Energy Statistics*, national renewables statistics, Department of Energy and Climate Change, London: The Stationery Office, 2011. The 2010 figure is explained as being due to low wind, and it is interesting that it does not figure in the highlighted numbers. See also IPCC, 'IPCC Special Report on Renewable Energy Sources and Climate Change Mitigation', prepared by Working Group III of the Intergovernmental Panel on Climate Change (Edenhofer, O., Pichs-Madruga, R., Sokona, Y., Seyboth, K., Matschoss, P., Kadner, S., Zwickel, T., Eickemeier, P., Hansen, G., Schlömer, S. and von Stechow, C. (eds)), Cambridge: Cambridge University Press, and New York, 2011.

4. Alex Salmond speaking during a visit to Steel Engineering in Renfrew, 27 April 2011. Reported by the *Daily Record*.
5. There is much dispute about how much wind creates this system problem. Some argue that conventional systems can absorb up to 20% wind without stability problems; others argue that it is much less.
6. European Commission, *Energy Roadmap 2050*, COM(2011)885/2, 2011.
7. The US is in a somewhat different situation, with enormous geographical extremes, and with an incomplete federal network. Yet it cannot escape the costs. A recent study suggested that the additional costs of transmission lines were on average about 30%. Mills, A. and Wiser, R., 'The Cost of Transmission for Wind Energy: A Review of Transmission Planning Studies', U.S. Department of Energy, Washington, D.C., February 2009.
8. Hedegaard, C., 'The EU's Role in Fighting Climate Change', in Barysch, K. (ed), *Green, Safe, Cheap: Where Next for EU Energy Policy?*
9. Ibid.
10. Chris Huhne on a visit to Newcastle, September 30 2011. 'Energy secretary Chris Huhne has told the North East to learn to love "beautiful" wind turbines and called for hundreds more to be given the go ahead', *The Journal*, 1 October 2011. See also Jonathon Porritt in *Country Life*, April 2004: 'I find wind turbines objects of compelling beauty.'
11. MacKay, D. J. C., *Sustainable Energy – Without the Hot Air*, Cambridge: UIT, 2008. In the case of the Scottish First Minister's 2012 ambition to reach 100% renewables by 2020 it has been estimated that to meet this objective by onshore wind would require 5000 wind turbines per conventional station replaced. There are three, and hence the land area required would be equivalent to eight times the size of Glasgow. See 'Interesting Figures' at www.esru.strath.ac.uk/EandE/Web_sites/01-02/RE_info/interesting.htm
12. MacKay, D. J. C., *Sustainable Energy – Without the Hot* Air, 2008, p. 33.
13. Ibid., p. 33.
14. Greenacre, P., Gross, R. and Heptonstall, P., 'Great Expectations: The Cost of Offshore Wind in UK waters – Understanding the Past and Projecting the Future', Technology and Policy Assessment Function of the UK Energy Research Centre, September 2010.
15. See also IPCC, 'Renewable Energy Sources and Climate Change Mitigation: Special Report of the Intergovernmental Panel on Climate Change', 2012.
16. MacKay, D. J. C., *Sustainable Energy – Without the Hot Air*, 2008, p. 61.
17. See Borenstein, S., 'The Private and Public Economics of Renewable Electricity Generation', 2012, for an explanation of levelized costs.
18. For a sample, see Committee on Climate Change and Mott MacDonald, 'Costs of Low Carbon Generation Technologies', 2011 and Department for Energy and Climate Change, 'Electricity Generation Cost Model – 2011', Update Revision 1, August 2011, Prepared by Parsons Brinckerhof.
19. Committee on Climate Change, 'The Renewable Energy Review', May 2011.
20. See the DESERTEC website: www.desertec.org
21. Germany reduced its subsidy by 16% in 2010, with annual reductions going forward depending on take-up. Spain capped the number of hours for subsidy in communal projects. Schemes have faced subsidy cuts in the Czech Republic and France, with confusion in Italy. In Britain cuts were applied from the middle of 2011 for larger schemes, and later for smaller ones. On the costs in Germany, see Frondel, M., Ritter, N., Schmidt, C. and Vance, C., 'Economic Impacts from the Promotion of Renewable Energy Technologies: The German Experience', Ruhr Economic Papers, no. 156, 2009.
22. Ibid.
23. Directive 2009/28/EC of the European Parliament and of the Council of 23 April 2009.
24. See, for example, the Euractiv.com report, 'Biomass Insanity May Threaten EU Carbon Targets', 2 April 2012, for differing views. Available at http://www.euractiv.com/energy/biomass-insanity-may-threaten-eu-carbon-targets-news-511891

25. See 'Biomass Program' on the US Department of Energy website at www1.eere.energy. gov/biomass. For a critique see Hahn, R. and Cecot, C., 'The Benefits and Costs of Ethanol: An Evaluation of the Government's Analysis', *Journal of Regulatory Economics*, 35(3), pp. 275–295. See also Koplow, D., 'Biofuels: At What Cost?', Global Subsidies Initiative, www.globalsubsidies.org

26. National Energy Policy Development Group, 'National Energy Policy: Reliable, Affordable, and Environmentally Sound Energy for America's Future', May 2001.

27. MacKay, D.J.C., *Sustainable Energy – Without the Hot Air*, 2008, p. 44.

28. Knittel, C., 'Reducing Petroleum Consumption from Transportation', *Journal of Economic Literature*, 26(1), Winter, 2012, pp. 93–118. See in particular pp. 100–104.

29. Commission of the European Communities, '20 20 by 2020: Europe's Climate Change Opportunity', Brussels, 23 January, COM (2008) 30 Final.

30. Ibid.

31. The definition from the Directive is: ' "energy from renewable sources" means energy from renewable non-fossil sources, namely wind, solar, aerothermal, geothermal, hydrothermal and ocean energy, hydropower, biomass, landfill gas, sewage treatment plant gas and biogases'.

32. The Crown Estate (2007), 'Supporting Information Relating to the Announcement of Round 3', at www.thecrownestate.co.uk/media/214342/round3_briefing_note.pdf

33. The minister responsibly reduced the expectation to around 13 GW for offshore wind in February 2011. See www.publications.parliament.uk/pa/cm201011/cmhansrd/cm110210/debtext/110210-0001.htm#qn_o6. It has fallen further since.

34. The exact numbers are subject to major dispute – not least because the costs vary from site to site, and depending on counterfactual assumptions. A sample of estimates is provided by the Committee on Climate Change report on renewables: Committee on Climate Change, 'The Renewable Energy Review', May 2011; House of Lords Select Committee on Economic Affairs, 'The Economics of Renewable Energy', 4th Report of Session 2007–08, HL Paper 195–I, London: The Stationery Office; and an analysis by the Renewable Energy Foundation, 'The Probable Cost of UK Renewable Electricity Subsidies 2002–2030', June 2011.

35. See Sinn, H-W., *The Green Paradox: A Supply-side Approach to Global Warming*, Cambridge, Massachusetts: MIT Press, 2012.

36. Ibid., Chapter 2.

37. Weyler, R., *Greenpeace: An Insider's Account*, London: Rodale, 2004.

38. An example of this is the Centre for Alternative Technology in Wales.

39. Schumacher, E. F., *Small is Beautiful: Economics as if People Mattered*, London: Blond & Briggs, 1973.

40. 'No generation has a freehold on this earth. All we have is a life tenancy – with a full repairing lease. This Government intends to meet the terms of that lease in full', speech to the Conservative Party Conference, 14 October 1988. Quoted in Harris, R., *The Collected Speeches of Margaret Thatcher*, London: HarperCollins, 1997.

41. Bosman, R., 'How Germany's Powerful Renewables Advocacy Coalition is Transforming the German (and European) Energy Market', *European Energy Review*, 27 February 2012. See also Siemens's website: www.siemens.com/press/en/feature/2012/corporate/2012-03-energiewende.php

Chapter 5: Can demand be cut?

1. Speech to the Liberal Democrat Autumn Conference, September 2011. See also Department of Energy and Climate Change, 'Estimated Impacts of Energy and Climate Change Policies on Energy Prices and Bills, November 2011.

2. Jevons, W. S., *The Coal Question: An Inquiry Concerning the Progress of the Nation, and the Probable Exhaustion of Our Coal-mines*, 2008 edition, p. 75.

3. For a survey, see Sorrell, S., 'Improving Energy Efficiency: Hidden Costs and Unintended Consequences', in Helm, D. and Hepburn, C., *The Economics and Politics of Climate Change*, Oxford: Oxford University Press, 2009. Sorrell also points out that the Stern Review manages to overlook the rebound effect altogether, despite relying heavily on energy efficiency to generate its cost estimates for climate change mitigation.
4. McKinsey Global Energy and Materials, 'Unlocking Energy Efficiency in the US Economy', 2009.
5. For a devastating critique of the methodology underlying these sorts of studies, and McKinsey's estimates in particular, see Allcott, H. and Greenstone, M., 'Is There an Energy Efficiency Gap?', *Journal of Economic Perspectives*, 26(1), 2012, pp. 3–28.
6. Stern, N., *A Blueprint for a Safer Planet: How to Manage Climate Change and Create a New Era of Progress and Prosperity*, London: Random House, 2009. See pp. 48–53 in particular.
7. See Allcott, H. and Greenstone, M., 'Is there an Energy Efficiency Gap?', 2012, and Hills, J., 'Getting the Measure of Fuel Poverty: Final Report of the Fuel Poverty Review', Centre of Analysis and Social Exclusion Report 72 commissioned by the Department of Energy and Climate Change, 15 March 2012.
8. For an excellent survey, see Kahneman, D., *Thinking, Fast and Slow*, London: Allen Lane, 2011.
9. Hamilton, J., 'Understanding Crude Oil Prices', *Energy Journal*, 30(2), 2009, pp. 179–206.
10. See Allcott, H. and Greenstone, M., 'Is there an Energy Efficiency Gap?', 2012, Figure 2, p. 7.
11. Wolfram, C., Shelef, O. and Gertler, P., 'How Will Energy Demand Develop in the Developing World?', *Journal of Economic Perspectives*, 26(1), 2012, pp. 119–138, provide a decomposition of demand for developing countries to substantiate this result.
12. See Allcott, H. and Greenstone, M., 'Is there an Energy Efficiency Gap?', 2012, p. 17.
13. For a summary of the Green Deal, see Richards, P., 'The Green Deal', Standard Note SN/SC/5763, House of Commons Library, 20 January 2012. Available at www.parliament.uk/briefing-papers/SN05763.pdf
14. For a general discussion of standards in environmental policy, see Anthoff, D. and Hahn, R., 'Government Failure and Market Failure: On the Inefficiency of Environmental and Energy Policy', *Oxford Review of Economic Policy*, 26(2), 2009, pp. 197–224.

Chapter 6: A new dawn for nuclear?

1. Contractually the Olkiluoto 3 reactor in Finland is being built for a consortium of industrial buyers.
2. See Helm, D., *Energy, the State and the Market: British Energy Policy Since 1979*, Oxford: Oxford University Press, 2004, Chapters 5 and 10.
3. Department of Trade and Industry, 'Our Energy Future: Creating a Low-carbon Economy', 2003.
4. Department for Business, Enterprise and Regulatory Reform, 'Meeting the Energy Challenge: A White Paper on Nuclear Power', HM Government, 2008.
5. This quote is taken from a video of Chris Huhne discussing nuclear power, which was available on the Liberal Democrats Party website. It was removed in October 2010.
6. In March 2012, E.ON and RWE, the two large German energy companies, withdrew from nuclear investment in Britain, leaving the majority French-state-owned EDF as the main champion of new build in Britain.
7. Translated from Polish Press Agency, interview with Donald Tusk, Prime Minister of Poland, 'Tusk: nie zmieniamy planów dot. energetyki jadrowej', 31st May 2011, Warsaw.

8. Curiously, Weyler fails to discuss this in his history of Greenpeace. Weyler, R., *Greenpeace: An Insider's Account*, 2004.

9. For example, see Saenko, V., Ivanov, V., Tsyb, A., Bogdanova, T., Tronko, M., Yu. D. and Yamashita, S., 'The Chernobyl Accident and its Consequences', *Clinical Oncology*, 23(4), May 2011, pp. 234–243.

10. Mönnig, A. and Wiebe, K., 'Macroeconomic Effects of the Current Crises in Japan and MENA Countries: A Model-based Assessment of the Medium Term', gws Discussion Paper 2011/1.

11. von Hippel, R. N., 'The Radiological and Psychological Consequences of the Fukushima Daiichi Accident', *Bulletin of the Atomic Scientists*, 67(5), September/October, 2011, pp. 27–36, estimates 1000 extra cancer deaths. More alarmingly, Cardis et al. (2006) estimate 14000 extra cancer deaths as a result of Chernobyl. Cardis, E., Howe, G., Ron, E., Bebeshko, V., Bogdanova, T., Bouville, A., Carr, Z., Chumak, V., Davis, S., Demidchik, Y., Drozdovitch, V., Gentner, N., Gudzenko, N., Hatch, M., Ivanov, V., Jacob, P., Kapitonova, E., Kenigsberg, Y., Kesminiene, A., Kopecky, K. J., Kryuchkov, V., Loos, A., Pinchera, A., Reiners, C., Repacholi, M., Shibata, Y., Shore, R. E., Thomas, G., Tirmarche, M., Yamashita, S. and Zvonova, I., 'Cancer Consequences of the Chernobyl Accident: 20 Years On', *Journal of Radiological Protection*, Institute of Physics Publishing, 26, 2006, pp. 127–140.

12. The Indian Ocean tsunami in 2004 killed an estimated 230,000 people. US Geological Survey, www.earthquake.usgs.gov

13. Flowers, B., 'Nuclear Power and the Environment', Sixth Report of the Royal Commission on Environmental Pollution, 1976.

14. United Kingdom Atomic Energy Authority and UK Nirex Ltd, Cmnd 884, 'The Control of Radioactive Wastes', London: HMSO, 1959.

15. A current example is the proposed PRISM reactor being promoted by GE Hitachi. See GE Hitachi Nuclear Energy, 'PRISM: Elegantly Simple, Passive, Modular, and Environmental', Technology Update, 2009.

16. Schelling, T. C., 'An Astonishing Sixty Years: The Legacy of Hiroshima', *American Economic Review*, American Economic Association, 2006, 96(4), pp. 929–937. See also Barrett, S., *Why Cooperate? The Incentive to Supply Global Public Goods*, 2007, for a discussion of the global public goods aspects of nuclear deterrence.

17. See Grubler, A., 'The Costs of the French Nuclear Scale-up: A Case of Negative Learning by Doing', *Energy Policy*, 38(9), 2010, pp. 5174–5188, and Davis, L. W., 'Prospects for Nuclear Power', *Journal of Economic Perspectives*, 26(1), 2012, pp. 49–66, 2012. Grubler points out that, 'In many ways, the French nuclear program was the ideal setting for encouraging learning-by-doing, so one might have expected costs to decrease over time' (p. 54).

18. Davis, L. W., 'Prospects for Nuclear Power', 2012.

Chapter 7: Are we running out of fossil fuels?

1. Hubbert, M. K., 'Nuclear Energy and the Fossil Fuels', *Drilling and Production Practice*, American Petroleum Institute & Shell Development Co. Publication No. 95, 1956.

2. The most vociferous was Colin Campbell. See for example, Campbell, C. and Laherrere, J., 'The End of Cheap Oil', *Scientific American*, March 1998. For a critique, see Maugeri, L., *The Age of Oil: The Mythology, History, and Future of the World's Most Controversial Resource*, London: Praeger, 2006.

3. Yergin, D., *The Prize: The Epic Quest for Oil, Money and Power*, New York: Free Press 1991.

4. Meadows, D. H., Meadows, D. L., Randers, J. and Behrens, W., 'The Limits to Growth: A Report for the Club of Rome's Project on the Predicament of Mankind', New York: Universe Books, 1972.

5. Simon, J. L., *The Ultimate Resource*, Oxford: Martin Robertson, 1981.

6. In 2011, the Norwegian company Statoil found one of the largest fields in the Norwegian continental shelf, Aldous, decades after oil was first found.
7. See Alex Kemp's official history of the British North Sea oil industry. Kemp, A., *The Official History of North Sea Oil and Gas*, 2 vols., Abingdon: Routledge, 2012.
8. Yergin, D., *The Quest: Energy, Security, and the Remaking of the Modern World*, London: Allan Lane, 2011, p. 262.
9. See again Yergin, D., *The Prize: The Epic Quest for Oil, Money and Power*, 1991.
10. See the EIA website on gas, and the energy briefing on shale gas. This also contains the early release 2012 forecasts for US gas production. www.eia.gov/forecasts/aeo/er/
11. There is a large literature on the sustainability of Chinese growth. See, for example, Whalley, J. and Xin, L., 'China's FDI and Non-FDI Economies and the Sustainability of Future High Chinese Growth', *China Economic Review*, 21(1), March 2010, pp. 123–135; and Chen, Y-J., 'An Endogenous Engine of Sustainable Growth for Chinese Economy: Review of 11th Five Year Plan and Prospects of the 12th Five year Plan for the Development of the Private Economy', *Economic Theory and Business Management*, 2, 2011.
12. Paul Stevens has written persuasively on these political constraints. See Stevens, P., *The Coming Oil Supply Crunch*, London: Chatham House, 2008.
13. On the oil case, see Rick van der Ploeg's excellent survey in 'Natural Resources, Curse or Blessing?', *Journal of Economic Literature*, vol. XLIX, no. 2, June 2011.
14. Small, K. A. and van Dender, K., 'Fuel Efficiency and Motor Vehicle Travel: The Declining Rebound Effect', Department of Economics, University of California, Irvine *Working Paper*, 2007; and Fontaras, G. and Samaras, Z., 'On the Way to 130g CO_2/km – Estimating the Future Characteristics of the Average European Passenger Car', *Energy Policy*, 38(4), April 2010, pp. 1826–1833.
15. An example of this alarmism is found in UK Industry Taskforce on Peak Oil and Energy Security, 'The Oil Crunch: A Wake-up Call for the UK Economy', Second Report of the UK Industry Taskforce on Peak Oil & Energy Security (ITPOES), February 2010. Interestingly it was led by three prominent industrialists, all with considerable interests in the results. As with the discussion of renewables, energy efficiency and nuclear lobbyists, separating out analysis from lobbying is far from straightforward when it comes to oil price alarmism.
16. Department of Energy and Climate Change, 'Davey: Climate Change Policies Could Halve Negative Impacts of Energy Price Shocks', press release 2012/061, 18 May 2012.
17. Department of Energy and Climate Change, 'Electricity Market Reform: Policy Overview', May 2012.
18. Lowe, P., 'The Future Fuel Mix for Europe', presentation by Philip Lowe, Director General for Energy, European Commission, to the FLAME Gas Conference, Amsterdam, 17 May 2012.

Chapter 8: Will there ever be a credible international agreement?

1. For the full text of the United Nations Framework Convention on Climate Change, see the UNFCCC website at http://unfccc.int/essential_background/convention/background/items/1349.php
2. Scott Barrett has led the way in explaining why this approach is so revealing in international public policy issues. See Barrett, S., 'Environment and Statecraft: The Strategy of Environmental Treaty-making', Oxford: Oxford University Press, 2005, for a comprehensive treatment; and also Barrett, S., *Why Cooperate? The Incentive to Supply Global Public Goods*, 2007, where he sets climate change alongside other global public goods problems.
3. Stern, N., 'Raising Consumption, Maintaining Growth and Reducing Emissions: The Objectives and Challenges of China's Radical Change in Strategy and its Implications for the World Economy', 2011.

4. Victor, D. G., *Global Warming Gridlock: Creating More Effective Strategies for Protecting the Planet*, Cambridge: Cambridge University Press, 2011. See also his earlier brilliant exposition of the flaws in Kyoto in Victor, D. G., *The Collapse of the Kyoto Protocol and the Struggle to Slow Global Warming*, Princeton: Princeton University Press, 2004.
5. Blair, T., *A Journey*, 2010.
6. European Environment Agency, 'The European Community's Initial Report Under the Kyoto Protocol Report to Facilitate the Calculation of the Assigned Amount of the European Community Pursuant to Article 3, Paragraphs 7 and 8 of the Kyoto Protocol: Submission to the UNFCCC Secretariat', EEA Technical Report No 10/2006.
7. FAO, UNDP, UNEP, 'UN Collaborative Programme on Reducing Emissions from Deforestation and Forest Degradation in Developing Countries: Framework Document', 20 June 2008.
8. Osborne, G., 'Together We Will Ride Out the Storm', speech to the Conservative Party Conference, 2 October 2011, available at: www.conservatives.com/news/speeches/2011/10/osborne_together_we_will_ride_out_the_storm.aspx
9. See UNFCCC website, 'Durban Climate Change Conference – November/December 2011', at http://unfccc.int/meetings/durban_nov_2011/meeting/6245.php

Chapter 9: Fixing the carbon price

1. Quoted in Vidal, J., Stratton, A. and Goldenberg, S., 'Low Targets, Goals Dropped: Copenhagen Ends in Failure', *The Guardian*, 19 December 2009.
2. For a critique of IAMs see DeCanio, S. J., *Economic Models of Climate Change*, London: Palgrave Macmillan, 2003.
3. This depends on how the income is treated and whether the permits are auctioned or grandfathered (given out).
4. For an excellent summary of the arguments see Hepburn, C., 'Regulating by Prices, Quantities or Both: An Update and an Overview', *Oxford Review of Economic Policy*, 22(2), 2006, pp. 226–247. For the classic treatment of the choice between prices and quantities see Weitzman, M. L., 'Prices vs Quantities', *Review of Economic Studies*, 41, 1974, pp. 477–491.
5. European Commission, 'Proposal for a Council Directive Introducing a Tax on Carbon Dioxide Emissions and Energy', COM (92) 226 final, 30 June 1992.
6. The forerunner of the EU ETS, the UK ETS, provided more than a free pass – it actually subsidized reductions in pollution that would probably have taken place anyway. Department of the Environment, Transport and the Regions, 'Climate Change: The UK Programme', Cm 4913, London, HMSO, 2000.
7. Quoted in Krukowska, E., 'EON's Teyssen Urges Fix to "Bust" EU CO_2 Plan, Energy Rules', Bloomberg, 7 February 2012.
8. Roberts, M. J. and Spence, M., 'Effluent Charges and Licenses under Uncertainty', *Journal of Public Economics*, 5, 1976, pp. 193–208. See also Helm, D. R., 'Caps and Floors for the EU ETS: A Practical Carbon Price', Polish Office of the Committee for European Integration, 4 September 2008.
9. HM Treasury and HM Revenue and Customs, 'Carbon Price Floor Consultation: The Government Response', London, The Stationery Office, 2011.
10. See Pearce, D., 'The Social Cost of Carbon', Mendelsohn, R., 'The Social Costs of Greenhouse Gases, their Values and Policy Implications', and Tol, R., 'The Marginal Damage Costs of Carbon Dioxide Emissions', in Helm, D. (ed), *Climate-change Policy*, Oxford: Oxford University Press, 2005.
11. See Helm, D., Hepburn C. and Marsh, C., 'Credible Carbon Taxes', in Helm, D. R. and Hepburn, C. (eds), *The Economics and Politics of Climate Change*, Oxford: Oxford University Press, 2011 edition.

258 NOTES TO PP. 192–222

258 NOTES TO PP. 192–222

12. For a game-theoretic model which captures the incentives on others to introduce their own internal carbon price, see Helm, D. R., Hepburn, C. and Ruta, G., 'Trade, Climate Change and the Political Game Theory of Border Carbon Adjustments', forthcoming in the *Oxford Review of Economic Policy*, Autumn, 2012.

Chapter 10: Making the transition

1. Helm, D. R., 'The Russian Dimension and Europe's External Energy Policy', September 2007. Available at www.dieterhelm.co.uk
2. The plan is for 6.5 billion cubic metres by 2015, and 60–100 billion cubic metres by 2020, from effectively zero in 2012. China National Development Reform Commission, 'Shale Gas Five Year Plan', March 2012.
3. See BBC News, 'Repsol Makes Big Shale Oil Find in Argentina', 8th November 2011, http://www.bbc.co.uk/news/business-15631423; energy-pedia News, 'YPF Makes New Shale Oil and Gas Find in Argentina', 30th March 2012, http://www.energy-pedia.com/news/argentina/ypf-makes-new-shale-oil-and-gas-find-in-argentina-149751; and *The Warsaw Voice*, 'Shale Gas: Poland's Big Chance?', April 2011, www.warsawvoice.pl/WVpage/pages/articlePA.php/23630/article
4. See Energy Information Administration, 'World Shale Gas Resources: An Initial Assessment of 14 Regions Outside the United States', 5 April 2011; World Energy Council, 'Survey of Energy Resources: Shale Gas: What's New?', 2012.
5. See, for example, WWF, 'Shale Gas Incompatible with Addressing Climate Change', 17 April 2012, www.wwf.org.uk/what_we_do/press_centre/?unewsid=5898; Greenpeace, 'Greenpeace Response to Government Gas Fracking Go-ahead', 17 April 2012, www.greenpeace.org.uk/media/press-releases/greenpeace-response-government-gas-fracking-go-ahead-20120417; and Friends of the Earth, 'Rush for Shale Gas Makes No Sense in Fight Against Climate Change', 24 May 2011, www.foe.co.uk/resource/press_releases/shale_gas_report_reaction_24052011.html
6. Green, C., Styles, P. and Baptie, B., 'Preese Hall Shale Gas Fracturing Review and Recommendations for Induced Seismic Mitigation', report for the Department of Energy and Climate Change, April 2012.
7. House of Commons Select Committee on Energy and Climate Change, 'Shale Gas: Fifth Report of Session 2010–2012', HC795, May 2011.
8. House of Commons Environmental Audit Committee, 'Carbon Capture and Storage: Ninth Report of Session 2007/2008', HC654, 2008. See also Hertzog, H., 'Carbon Dioxide Capture and Storage', in Helm, D. R. and Hepburn, C. (eds), *The Economics and Politics of Climate Change*, 2011 edition.
9. Committee on Climate Change, 'Unabated Gas-fired Generation', 24 May 2012, http://www.theccc.org.uk/news
10. For more information see www.bgs.ac.uk/discoveringGeology/climateChange/ccs/howcanCO2bestored.html
11. Energy Information Administration, 'Short-term Energy Outlook', June 2012.
12. US Energy Information Administration, 'Annual Energy Outlook 2012: Early Release'.
13. International Energy Agency, 'Global Carbon-dioxide Emissions Increase by 1.0 Gt in 2011 to Record High', May 2012. Available at: www.iea.org/newsroomandevents/news/2012/may/name,27216,en.html

Chapter 11: Investing in new technologies

1. Crown Estate, 'Offshore Wind Cost Reduction Pathways Study', 12 June 2012, London: The Stationery Office.
2. Haney, A., Jamasb, T. and Pollitt, M., 'Smart Metering and Electricity Demand: Technology, Economics and International Experience', EPRG Working Paper EPRG 0903, Cambridge Working Paper in Economics 0905, 2009. See also Venables, M.,

'Smart Meters Make Smart Consumers [Analysis]', *Engineering & Technology*, 2(4), April 2007.

3. Rio Tinto now uses driverless trucks (and driverless trains) in some of its mining operations in Australia. www.riotinto.com/media/5157_21165.asp

4. See the Joint Center for Artificial Photosynthesis. www.solarfuelshub.org

5. Shirvani, T., Inderwildi, O. R., Yan, X., Edwards, P. P. and King, D. A., 'Life Cycle Analysis and Greenhouse Gas Analysis for Algae-derived Biodiesel', *Energy & Environmental Science*, 2011 DOI: 10.10139/C1EE01791H.

6. A brief summary of the story of Harrison's clock can be found on the Royal Maritime Museum website, www.rmg.co.uk/harrison

7. For a critique of the competition, see National Audit Office, 'Carbon Capture and Storage: Lessons from the Competition for the First UK Demonstration', London, The Stationery Office, 2012.

8. Noll, R. and Cohen, L., *The Technology Pork Barrel*, Brookings Institute, Washington, 1991.

9. The Manhattan Project was the US R&D programme that led to the production of the first atomic bomb during the Second World War.

Conclusion

1. Wilson, E. O., *In Search of Nature*, London: Allen Lane, 1996, p. 184.

2. Department of Trade and Industry, 'Our Energy Future: Creating a Low-carbon Economy', Energy White Paper, Foreword.

3. Speech to the House of Commons by Prime Minister Winston Churchill, 13 May 1940.

Bibliography

Allcott, H. and Greenstone, M., 'Is there an Energy Efficiency Gap?', *Journal of Economic Perspectives*, 26(1), 2012, pp. 3–28

Anthoff, D. and Hahn, R., 'Government Failure and Market Failure: On the Inefficiency of Environmental and Energy Policy', *Oxford Review of Economic Policy*, 26(2), 2009, pp. 197–224

Anthoff, D. and Tol, R., 'The Uncertainty about the Social Cost of Carbon: A Decomposition Analysis Using FUND', ESRI Working Paper, 2011

Barrett, S., *Environment and Statecraft: The Strategy of Environmental Treaty-making*, Oxford: Oxford University Press, 2005

Barrett, S., *Why Cooperate? The Incentive to Supply Global Public Goods*, New York: Oxford University Press,, 2007

Barysch, K. (ed.), *Green, Safe, Cheap: Where Next for EU Energy Policy?*, Centre for Economic Reform, London, 2011

Blair, T., *A Journey*, London: Hutchinson, 2010

Böhringer, C., Carbone, J. C. and Rutherford, T. F., 'Embodied Carbon Tariffs', National Bureau of Economic Research, working paper series, 17376, August 2007

Borenstein, S., 'The Private and Public Economics of Renewable Electricity Generation', *Journal of Economic Perspectives*, American Economic Association, 26(1), 2012, pp. 67–92

Bosman, R., 'How Germany's Powerful Renewables Advocacy Coalition is Transforming the German (and European) Energy Market', *European Energy Review*, 27 February 2012

Campbell, C. and Laherrere, J., 'The End of Cheap Oil', *Scientific American*, March 1998

Cardis, E., Howe, G., Ron, E., Bebeshko, V., Bogdanova, T., Bouville, A., Carr, Z., Chumak, V., Davis, S., Demidchik, Y., Drozdovitch, V., Gentner, N., Gudzenko, N., Hatch, M., Ivanov, V., Jacob, P., Kapitonova, E., Kenigsberg, Y., Kesminiene, A., Kopecky, K. J., Kryuchkov, V., Loos, A., Pinchera, A., Reiners, C., Repacholi, M., Shibata, Y., Shore, R. E., Thomas, G., Tirmarche, M., Yamashita, S. and Zvonova, I., 'Cancer Consequences of the Chernobyl Accident: 20 Years On', *Journal of Radiological Protection*, Institute of Physics Publishing, 26, 2006, pp. 127–140

Chamon, M., Mauro, P. and Okawa, Y., 'Cars: Mass Ownership in the Emerging Market Giants, *Economic Policy*, 23(54), 2008, pp. 243–296

Chen, Y-J., 'An Endogenous Engine of Sustainable Growth for Chinese Economy: Review of 11th Five Year Plan and Prospects of the 12th Five year Plan for the Development of the Private Economy', *Economic Theory and Business Management*, 2, 2011

China National Development Reform Commission, 'Shale Gas Five Year Plan', March 2012

Coase, R. H., 'The Problem of Social Cost', *Journal of Law and Economics*, 3, 1960, pp. 1–44

Collier, P., *The Bottom Billion: Why the Poorest Countries are Failing and What Can be Done About it*, New York: Oxford University Press, 2007

Collier, P., Conway, G. and Venables, T., 'Climate change and Africa', Chapter 7 in Helm, D. and Hepburn, C. (eds), *The Economics and Politics of Climate Change*, Oxford: Oxford University Press, 2009

Commission of the European Communities, '20 20 by 2020: Europe's Climate Change Opportunity', Brussels, 23 January, COM (2008) 30 Final

Committee on Climate Change, 'The Renewable Energy Review', London: The Stationery Office, May 2011

Committee on Climate Change and Mott MacDonald, *Costs of Low Carbon Generation Technologies*, London: The Stationery Office, 2011

Crawford, R. M., *Plants at the Margin: Ecological Limits and Climate Change*, Cambridge: Cambridge University Press, 2009

Crown Estate, 'Offshore Wind Cost Reduction Pathways Study', London: The Stationery Office, 12 June 2012

D'Andrea, W. J., Huang, Y., Fritz, F. C. and Anderson, N. J., 'Abrupt Holocene Climate Change as an Important Factor for Human Migration in West Greenland', Proceedings of the National Academy of Sciences of the United States of America, 2011

Davis, L. W., 'Prospects for Nuclear Power', *Journal of Economic Perspectives*, 26(1), 2012, pp. 49–66

DeCanio, S. J., *Economic Models of Climate Change*, London: Palgrave Macmillan, 2003

Department for Business, Enterprise and Regulatory Reform, 'Meeting the Energy Challenge: A White Paper on Nuclear Power', HM Government, 2008

Department of Energy and Climate Change, *Digest of United Kingdom Energy Statistics 2011*, London: The Stationery Office, 2011

Department of Energy and Climate Change, 'UK Renewable Energy Roadmap', July 2011

Department of Energy and Climate Change, *Electricity Generation Cost Model – 2011* Update Revision 1, August 2011, prepared by Parsons Brinckerhof

Department of Energy and Climate Change, 'Estimated Impacts of Energy and Climate Change Policies on Energy Prices and Bills', November 2011

Department of Energy and Climate Change, 'Electricity Market Reform: Policy Overview', May 2012

Department for Environment, Food and Rural Affairs, 'Valuing the Social Cost of Carbon Emissions: DEFRA Guidance', HM Government, 2002

Department for Environment, Transport and the Regions, 'Climate Change: The UK Programme', Cm 4913', London, HMSO, 2000

Department of Energy and Climate Change, 'Davey: Climate Change Policies Could Halve Negative Impacts of Energy Price Shocks', press release 2012/061, 18 May 2012

Department of Trade and Industry, 'Our Energy Future: Creating a Low-carbon Economy', Energy White Paper, The Stationery Office, 2003

Energy Information Administration, 'World Shale Gas Resources: An Initial Assessment of 14 Regions Outside the United States', 5 April 2011

Energy Information Administration, 'Short-term Energy Outlook', June 2012

European Commission, 'Proposal for a Council Directive Introducing a Tax on Carbon Dioxide Emissions and Energy', COM (92) 226 final, 30 June 1992

European Commission, 'Achieving a Deal on Climate Change: An EU View on Copenhagen Council of Foreign Relations New York', speech by José Manuel Durão Barroso, President of the European Commission, 21 September 2009

European Commission, *Energy Roadmap 2050* COM(2011)885/2, 2011

European Environment Agency, 'The European Community's Initial Report Under the Kyoto Protocol Report to Facilitate the Calculation of the Assigned Amount of the European Community Pursuant to Article 3, Paragraphs 7 and 8 of the Kyoto Protocol: Submission to the UNFCCC Secretariat', EEA Technical Report No 10/2006

Fagan, B., *The Long Summer: How Climate Changed Civilization*, London: Granta, 2004

FAO, UNDP, UNEP, 'UN Collaborative Programme on Reducing Emissions from Deforestation and Forest Degradation in Developing Countries: Framework Document', 20 June 2008

Flowers, B., 'Nuclear Power and the Environment', Sixth Report of the Royal Commission on Environmental Pollution, 1976

Fontaras, G. and Samaras, Z., 'On the Way to 130g CO_2/km – Estimating the Future Characteristics of the Average European Passenger Car', *Energy Policy*, 38(4), April 2010, pp. 1826–1833

Frondel, M., Ritter, N., Schmidt, C. and Vance, C. 'Economic Impacts from the Promotion of Renewable Energy Technologies: The German Experience', Ruhr Economic Papers, no. 156, 2009

GE Hitachi Nuclear Energy, 'PRISM: Elegantly Simple, Passive, Modular, and Environmental', Technology Update, 2009

Ghose, M. K. and Majee, S. R., 'Air Pollution Caused by Opencast Mining and its Abatement Measures in India', *Journal of Environmental Management*, 63(2), October 2001, pp. 193–202

Green, C., Styles, P. and Baptie, B., 'Preese Hall Shale Gas Fracturing Review and Recommendations for Induced Seismic Mitigation', report for the Department of Energy and Climate Change, London: The Stationery Office, April 2012

Greenacre, P., Gross, R. and Heptonstall, P., 'Great Expectations: The Cost of Offshore Wind in UK Waters – Understanding the Past and Projecting the Future', Technology and Policy Assessment Function of the UK Energy Research Centre, September 2010

Greenpeace, 'Greenpeace Response to Government Gas Fracking Go-ahead', 17 April 2012, www.greenpeace.org.uk/media/press-releases/greenpeace-response-government-gas-fracking-go-ahead-20120417

Greenstone, M. and Looney, A., 'Paying Too Much for Energy? The True Costs of Our Energy Choices', Working paper 12-05, MIT Department of Economics, February 2012

Grubler, A., 'The Costs of the French Nuclear Scale-up: A Case of Negative Learning by Doing', *Energy Policy*, 38(9), 2010, pp. 5174–5188

Hahn, R. and Cecot, C., 'The Benefits and Costs of Ethanol: An Evaluation of the Government's Analysis', *Journal of Regulatory Economics*, 35(3), pp. 275–295

Hamilton, J., 'Understanding Crude Oil Prices', *Energy Journal*, 30(2), 2009, pp. 179–206

Haney, A., Jamasb, T. and Pollitt, M., 'Smart Metering and Electricity Demand: Technology, Economics and International Experience', EPRG Working Paper EPRG 0903, Cambridge Working Paper in Economics 0905, 2009

Hardin, G., 'The Tragedy of the Commons', *Science*, 162, 1968, pp. 1243–1248

Hardy, J. T., *Climate Change: Causes, Effects and Solutions*, Chichester: J Wiley, 2003

Harris, R., *The Collected Speeches of Margaret Thatcher*, London: HarperCollins, 1997

Hedegaard, C., 'The EU's Role in Fighting Climate Change', in Barysch, K. (ed.), *Green, Safe, Cheap: Where Next for EU Energy Policy?*, London: Centre for Economic Reform, 2011

Helm, D., *Energy, the State and the Market: British Energy Policy Since 1979*, Oxford: Oxford University Press, 2004

Helm, D. (ed), *Climate-change Policy*, Oxford: Oxford University Press, 2005

Helm, D. R., 'The Russian Dimension and Europe's External Energy Policy', September 2007, www.dieterhelm.co.uk

Helm, D. R., 'Caps and Floors for the EU ETS: A Practical Carbon Price', Polish Office of the Committee for European Integration, 4 September 2008, www.dieterhelm.co.uk

Helm, D. R., 'Infrastructure, Investment and the Economic Crisis', in Helm, D., Wardlaw, J. and Caldecott, B., Delivering a 21st Century Infrastructure for Britain, Policy Exchange, 2009

Helm, D. R., 'Sustainable Consumption, Climate Change and Future Generations', The Royal Institute of Philosophy Supplements, 69(1), 2011, pp. 235–252

Helm, D. R. and Hepburn, C. (eds), The Economics and Politics of Climate Change, Oxford: Oxford University Press, 2011 edition

Helm, D., Hepburn C. and Marsh, C., 'Credible Carbon Taxes', in Helm, D. R. and Hepburn, C. (eds), The Economics and Politics of Climate Change, Oxford: Oxford University Press, 2011 edition

Helm, D. R., Hepburn, C. and Ruta, G., 'Trade, Climate Change and the Political Game Theory of Border Carbon Adjustments', forthcoming in the Oxford Review of Economic Policy, Autumn 2012

Helm, D., Phillips, J. and Smale, R., 'Too Good to be True? The UK's Climate Change Record', 2008, www.dieterhelm.co.uk

Hepburn, C., 'Regulating by Prices, Quantities or Both: An Update and an Overview', Oxford Review of Economic Policy, 22(2), 2006, pp. 226–247

Hertzog, H., 'Carbon Dioxide Capture and Storage', in Helm, D. R. and Hepburn, C. (eds), The Economics and Politics of Climate Change, Oxford: Oxford University Press, 2011 edition

Hillman, M., How We Can Save the Planet, London: Penguin Books, 2004

Hills, J., 'Getting the Measure of Fuel Poverty: Final Report of the Fuel Poverty Review', Centre of Analysis and Social Exclusion Report 72 commissioned by the Department of Energy and Climate Change, 15 March 2012

HM Treasury and HM Revenue and Customs, 'Carbon Price Floor Consultation: The Government Response', London: The Stationery Office, 2011

House of Commons Environmental Audit Committee, 'Carbon Capture and Storage: Ninth Report of Session 2007/2008', HC654, 2008

House of Commons Select Committee on Energy and Climate Change, 'Shale Gas: Fifth Report of Session 2010–2012', HC795, May 2011

House of Lords Select Committee on Economic Affairs, The Economics of Renewable Energy, 4th Report of Session 2007–08, HL Paper 195-I, London: The Stationery Office, 2008

Hubbert, M. K., 'Nuclear Energy and the Fossil Fuels', Drilling and Production Practice, American Petroleum Institute & Shell Development Co. Publication No. 95, 1956

Hume, D., Enquiries Concerning Human Understanding and Concerning the Principles of Morals, reprinted from the 1777 edition with an introduction by L. A. Selby-Bigge, Third Edition, Oxford: Oxford University Press, 1975

Hvistendahl, M., 'Coal Ash is More Radioactive than Nuclear Waste', Scientific American, 13 December 2007

International Energy Agency, 'CO_2 Emissions from Fuel Combustion: 2011 Highlights'

International Energy Agency, 'Coal Information, 2011', 2011

International Energy Agency, 'Coal: Medium-term Market Report: Market Trends and Projections to 2016', December 2011

International Energy Agency, World Energy Outlook 2011, Paris: OECD, 2011

International Energy Agency, 'Global Carbon-dioxide Emissions Increase by 1.0 Gt in 2011 to Record High', May 2012, www.iea.org/newsroomandevents/news/2012/may/name,27216,en.html

IPCC, 'Aviation and the Global Atmosphere: A Special Report of IPCC Working Groups I and III', 1999

IPCC, 'Impacts, Adaptation and Vulnerability: Contribution of Working Group II to the Fourth Assessment Report of the Intergovernmental Panel on Climate Change', IPCC AR4 2007 Working Group Report, Parry, M. L., Canziani, O. F., Palutikof, J. P., van der Linden, P. J. and Hanson, C. E. (eds), Cambridge: Cambridge University Press, 2007

IPCC, 'IPCC Special Report on Renewable Energy Sources and Climate Change Mitigation', prepared by Working Group III of the Intergovernmental Panel on Climate Change, Edenhofer, O., Pichs-Madruga, R., Sokona, Y., Seyboth, K., Matschoss, P., Kadner, S., Zwickel, T., Eickemeier, P., Hansen, G., Schlömer, S. and von Stechow, C. (eds), Cambridge: Cambridge University Press, and New York, 2011

IPCC, 'Renewable Energy Sources and Climate Change Mitigation: Special Report of the Intergovernmental Panel on Climate Change', Cambridge: Cambridge University Press, 2012

Jevons, W. S., *The Coal Question: An Inquiry Concerning the Progress of the Nation, and the Probable Exhaustion of Our Coal-mines*, London: Dodo Press, 2008 edition

Kahneman, D., *Thinking, Fast and Slow*, London: Allen Lane, 2011

Kemp, A., *The Official History of North Sea Oil and Gas*, 2 vols., Abingdon: Routledge, 2012

Knittel, C., 'Reducing Petroleum Consumption from Transportation', *Journal of Economic Literature*, 26(1), Winter, 2012, pp. 93–118. See in particular pp. 100–104

Koplow, D., 'Biofuels: At What Cost?', Global Subsidies Initiative, www.globalsubsidies.org

Krukowska, E., 'EON's Teyssen Urges Fix to "Bust" EU CO2 Plan, Energy Rules', Bloomberg, 7 February 2012

Lamb, H., *Climate, History and the Modern World*, 2nd edn, London: Routledge, 1995

Lane, N., *Life Ascending: The Ten Great Inventions of Evolution*, London: Profile Books, 2010

Lawson, N., *An Appeal to Reason: A Cool Look at Global Warming*, London: Duckworth, 2008

Lomborg, B., *The Skeptical Environmentalist: Measuring the Real State of the World*, Cambridge: Cambridge University Press, 1998

Lowe, P., *The Future Fuel Mix for Europe*, presentation by Philip Lowe, Director General for Energy, European Commission, to the FLAME Gas Conference, Amsterdam, 17 May 2012

MacKay, D. J. C., *Sustainable Energy – Without the Hot Air*, Cambridge: UIT, 2008

Marshall Task Force, 'Economic Instruments and the Business Use of Energy: Conclusions', Marshall Task Force on the Industrial Use of Energy, November 1998

Maugeri, L., *The Age of Oil: The Mythology, History, and Future of the World's Most Controversial Resource*, London: Praeger, 2006

McKinsey Global Energy and Materials, 'Unlocking Energy Efficiency in the US Economy', 2009

Meadows, D. H., Meadows, D. L., Randers, J. and Behrens, W., 'The Limits to Growth: A Report for the Club of Rome's Project on the Predicament of Mankind', New York: Universe Books, 1972

Mendelsohn, R., 'The Social Costs of Greenhouse Gases, their Values and Policy Implications', in Helm, D. (ed), *Climate-change Policy*, Oxford: Oxford University Press, 2005

Miller, G. H., et al., 'Abrupt Onset of the Little Ice Age Triggered by Volcanism and Sustained by Sea-ice/Ocean Feedbacks', *Geophysical Research Letters*, 39, L02708, 5 PP., 2012

Miller, R. W., *Globalizing Justice: The Ethics of Poverty and Power*, Oxford: Oxford University Press, 2010

Mills, A. and Wiser, R., 'The Cost of Transmission for Wind Energy: A Review of Transmission Planning Studies', U.S. Department of Energy, Washington, D.C., February 2009

Monbiot, G., *Heat: How to Stop the Planet Burning*, London: Penguin Books, 2007

Mönnig, A. and Wiebe, K., 'Macroeconomic Effects of the Current Crises in Japan and MENA Countries: A Model-based Assessment of the Medium Term', gws Discussion Paper 2011/1

National Academy of Sciences, *The Hidden Costs of Energy*, National Academics Press, 2010

National Audit Office, 'Carbon Capture and Storage: Lessons from the Competition for the First UK Demonstration', London: The Stationery Office, 2012

National Energy Policy Development Group, 'National Energy Policy: Reliable, Affordable, and Environmentally Sound Energy for America's Future', May 2001

Noll, R. and Cohen, L., *The Technology Pork Barrel*, Brookings Institute, Washington, 1991

Nordhaus, W., *A Question of Balance: Weighing the Options on Global Warming Policies*, New Haven: Yale University Press, 2008

Nozick, R., *Anarchy, State and Utopia*, Oxford: Blackwell, 1974

OECD and International Energy Agency, 'Medium-term Coal Market Report 2011: Market Trends and Projections', OECD/International Energy Agency, 2011

Osborne, G., 'Together We Will Ride Out the Storm', speech to the Conservative Party Conference, 2 October 2011, www.conservatives.com/news/speeches/2011/10/osborne_together_we_will_ride_out_the_storm.aspx

Pearce, D., 'Environmental Appraisal and Environmental Policy in the European Union', *Environmental and Resource Economics*, 11(3–4), 1998, pp. 489–501

Pearce, D., 'The Social Cost of Carbon', in Helm, D. (ed.), *Climate-change Policy*, Oxford: Oxford University Press, 2005

Popper, K., *The Logic of Scientific Discovery*, Vienna: Verlag von Julius Springer, 1935

Popper, K., *Conjectures and Refutations: The Growth of Scientific Knowledge*, London: Routledge, 1963

Posner, R. A., *Catastrophe: Risk and Response*, New York: Oxford University Press, 2004

Radioactive Substances Advisory Committee, Panel on Disposal of Radioactive Wastes Cmnd 884, 'The Control of Radioactive Wastes', London: HMSO, 1959

Ramsey, F., 'A Mathematical Theory of Saving', *Economic Journal*, 38(152), 1928, pp. 543–559

Rawls, J., *A Theory of Justice*, Cambridge, Massachusetts: Harvard University Press, 1971

Rees, M., *Our Final Century: Will the Human Race Survive the Twenty-first Century?*, London: Heinemann, 2003

Rees, M., *Our Final Hour*, New York: Basic Books, 2003

Reinhart, C. M. and Rogoff, K. S., *This Time is Different: Eight Centuries of Financial Folly*, New Jersey: Princeton University Press, 2009

Renewable Energy Foundation, *The Probable Cost of UK Renewable Electricity Subsidies 2002–2030*, June 2011

Richards, P., 'The Green Deal', Standard Note SN/SC/5763, House of Commons Library, 20 January 2012, www.parliament.uk/briefing-papers/SN05763.pdf

Roberts, M. J. and Spence, M., 'Effluent Charges and Licenses under Uncertainty', *Journal of Public Economics*, 5, 1976, pp. 193–208

Saenko, V., Ivanov, V., Tsyb, A., Bogdanova, T., Tronko, M., Yu. D. and Yamashita, S., 'The Chernobyl Accident and its Consequences', *Clinical Oncology*, 23(4), May 2011, pp. 234–243

Schelling, T. C., 'An Astonishing Sixty Years: The Legacy of Hiroshima', *American Economic Review*, American Economic Association, 2006, 96(4), pp. 929–937

Schumacher, E. F., *Small is Beautiful: Economics as if People Mattered*, London: Blond & Briggs, 1973

Scruton, R., *Green Philosophy: How to Think Seriously About the Planet*, London: Atlantic Books, 2012

Sen, A., 'Adam Smith and the Contemporary World', *Erasmus Journal for Philosophy and Economics*, 3(1), 2010, pp. 50–67

Shirvani, T., Inderwildi, O. R., Yan, X., Edwards, P. P. and King, D. A., 'Life Cycle Analysis and Greenhouse Gas Analysis for Algae-derived Biodiesel', *Energy & Environmental Science*, 2011 DOI: 10.10139/C1EE01791

Simon, J. L., *The Ultimate Resource*, Oxford: Martin Robertson, 1981

Sinn, H-W., *The Green Paradox: A Supply-side Approach to Global Warming*, Cambridge, Massachusetts: MIT Press, 2012

Small, K. A. and van Dender, K., 'Fuel Efficiency and Motor Vehicle Travel: The Declining Rebound Effect', Department of Economics, University of California, Irvine Working Paper, 2007

Sorrell, S., 'Improving Energy Efficiency: Hidden Costs and Unintended Consequences', in Helm, D. and Hepburn, C., *The Economics and Politics of Climate Change*, Oxford: Oxford University Press, 2009

Stern, N., 'The Economics of Climate Change: The Stern Review', HM Treasury, Cambridge: Cambridge University Press, January 2007

Stern, N., *A Blueprint for a Safer Planet: How to Manage Climate Change and Create a New Era of Progress and Prosperity*, London: Random House, 2009

Stern, N., 'Raising Consumption, Maintaining Growth and Reducing Emissions: The Objectives and Challenges of China's Radical Change in Strategy and its Implications for the World Economy', *World Economics*, 12(4), 2011, pp. 13–34

Stevens, P., *The Coming Oil Supply Crunch*, London: Chatham House, 2008

Stott, P. A, Stone, D. A. and Allen, M. R., 'Human Contribution to the European Heatwave of 2003', *Nature*, 432(2), December 2004, pp. 610–614

Tol, R., 'The Marginal Damage Costs of Carbon Dioxide Emissions', in Helm, D. (ed), *Climate-change Policy*, Oxford: Oxford University Press, 2005

Tol, R. 'The Economic Effects of Climate Change', *Journal of Economic Perspectives*, 33(2) Spring 2009, pp. 29–51

UK Industry Taskforce on Peak Oil and Energy Security, 'The Oil Crunch: A Wake-up Call for the UK Economy', Second Report of the UK Industry Taskforce on Peak Oil & Energy Security (ITPOES), February 2010

United Nations, 'The World at Six Billion', Department of Economic and Social Affairs, Population Division, October 1999

US Energy Information Administration, *International Energy Statistics 2011*, Washington DC: EIA, 2011

US Geological Survey, '90 Billion Barrels of Oil and 1,670 Trillion Cubic Feet of Natural Gas Assessed in the Arctic', news release, 23 July 2008, www.usgs.gov/newsroom/article. asp?ID=1980&from=rss_home

Vahrenholt, F., *Die Kalte Sonne*, Hoffmann & Campe Vlg Gmbh, February 2012

van der Ploeg, R., 'Natural Resources, Curse or Blessing?', *Journal of Economic Literature*, vol. XLIX, no. 2, June 2011

Venables, M., 'Smart Meters Make Smart Consumers [Analysis]', *Engineering & Technology*, 2(4), April 2007

Victor, D. G., *The Collapse of the Kyoto Protocol and the Struggle to Slow Global Warming*, Princeton: Princeton University Press, 2004

Victor, D. G., *Global Warming Gridlock: Creating More Effective Strategies for Protecting the Planet*, Cambridge: Cambridge University Press, 2011

Vidal, J., Stratton, A. and Goldenberg, S., 'Low Targets, Goals Dropped: Copenhagen Ends in Failure', *Guardian*, 19 December 2009

von Hippel, R. N., 'The Radiological and Psychological Consequences of the Fukushima Daiichi Accident', *Bulletin of the Atomic Scientists*, 67(5), September/October 2011, p. 27–36

Weart, S. R., *The Discovery of Global Warming*, London: Harvard University Press, 2003

Weber, E. U. and Stern, P. C., 'Public Understanding of Climate Change in the United States', *American Psychologist*, 66(4), May–June 2011, pp. 315–328

Weitzman, M. L., 'Prices vs Quantities', *Review of Economic Studies*, 41, 1974, pp. 477–491

Weyler, R., *Greenpeace: An Insider's Account*, London: Rodale, 2004

Whalley, J. and Xin, L., 'China's FDI and Non-FDI Economies and the Sustainability of Future High Chinese Growth', *China Economic Review*, 21(1), March 2010, pp. 123–135

Wilson, E. O., *The Diversity of Life*, London: Penguin, 1994

Wilson, E. O., *In Search of Nature*, London: Allen Lane, 1996

Wolfram, C., Shelef, O. and Gertler, P., 'How Will Energy Demand Develop in the Developing World?', *Journal of Economic Perspectives*, 26(1), 2012, pp. 119–38

World Energy Council, 'Survey of Energy Resources: Shale Gas: What's New?', 2012

WWF, 'Shale Gas Incompatible with Addressing Climate Change', 17 April 2012, www.wwf.org.uk/what_we_do/press_centre/?unewsid=5898

Yale Project on Climate Change Communication, 'Americans' Global Warming Beliefs and Attitudes in November 2011', Center for Climate Change Communication, George Mason University, November 2011

Yale Project on Climate Change Communication, 'Public Support for Climate and Energy Policies in November 2011', Center for Climate Change Communication, George Mason University, November 2011

Yergin, D., *The Prize: The Epic Quest for Oil, Money and Power*, New York: Free Press, 1991

Yergin, D., *The Quest: Energy, Security, and the Remaking of the Modern World*, London: Allen Lane, 2011

Index